E. A. Glysson
'75

AF817298

CALVIN COLLEGE
ENGINEERING DEPT.

INCINERATION OF SOLID WASTES

INCINERATION OF SOLID WASTES

Fred N. Rubel

NOYES DATA CORPORATION
Park Ridge, New Jersey London, England
1974

Copyright © 1974 by Noyes Data Corporation
No part of this book may be reproduced in any form
without permission in writing from the Publisher.
Library of Congress Catalog Card Number: 74-77723
ISBN: 0-8155-0551-5
Printed in the United States

Published in the United States of America by
Noyes Data Corporation
Noyes Building, Park Ridge, New Jersey 07656

Foreword

This Pollution Technology Review gives advanced technical and economic information designed to promote efficient incineration of municipal refuse, and will help to erase the bad neighbor image of incinerator installations within the community. Special incinerator applications, largely outside the scope of municipal waste disposal, have not been neglected.

Incineration offers the most significant volume reduction of solid organic wastes when compared with all other disposal methods.

The most undesirable side effect is, or rather was, air pollution, and here the primary concern is with emissions of particulates rather than with gases and odors. Many sophisticated devices for nearly complete retention of these solid particles have been developed in recent years, and special emphasis has been placed upon them in the descriptions of apparatus.

Most of the data and other information presented in this book are based upon detailed reports of government contracts fulfilled by industrial companies under the auspices of the Environmental Protection Agency.

In the United States, we are fortunate in receiving direct help from the numerous surveys, together with active research and development programs that are being supported by the Federal Government to help industry and municipalities to control their wastes and rubbish.

In this book are condensed vital data that are scattered and difficult to pull together. Important processes are interpreted and explained by actual case histories. This condensed information will provide a sound background for action towards combating pollution by solid waste management through incineration.

Advanced composition and production methods developed by Noyes Data are employed to bring our new durably bound books to the reader in a minimum of time. Special techniques are used to close the gap between "manuscript" and "completed book." Industrial technology is progressing so rapidly that time-honored, conventional typesetting, binding and shipping methods are no longer suitable. Delays in the conventional book publishing cycle have been bypassed to provide the user with an effective and convenient means of reviewing up-to-date information in depth.

The Table of Contents is organized in such a way as to serve as a subject index and provides easy access to the information contained in this book.

Contents and Subject Index

Introduction

The present work is a compilation of federally sponsored studies and reports in the field of incineration. In bringing them together, particularly theoretical topics concerning incineration have been avoided in favor of such fundamental matters as overall design, costs, emissions, and emission controls, all of which are presented in Part I of the book.

For a theoretical treatment of batch incinerator variables and design, the reader is referred to a recent two volume report entitled *Design and Control of Incinerators,* PB 223,626 September 1973, prepared by the Massachusetts Institute of Technology for the U.S. Environmental Protection Agency.

Part II of this book presents experimental and recent applications of incineration which may point to possibilities of utilizing incineration as a solution to some particularly difficult waste disposal problems.

As pressure increases to develop less polluting waste disposal techniques in the face of a worsening solid waste problem, so does the need for improved incinerator performance. This calls for a greater understanding of incineration as a technique. This applies particularly to municipalities which are considering the installation of an incinerator, to incinerator facility managers and workers, and of course to design engineers.

Pertinent portions of seventeen government publications on incineration, some of which are the result of a review of a large number of related publications in this field, are presented here under one cover, in a coherent approach to the subject. The government publications referred to are the following:

PB 223,626 1973
PB 200,514 June 1971

PB 217,821 (Public Health Service Publication No. 2032) 1970
PB 211,264 July 1971
PB 190,249 November 1969
PB 216,372 1970
PB 210,382 February 1972
PB 215,093 1965
PB 223,626 September 1973
PB 214,882 June 1958
PB 213,378 1972
PB 197,888 April 1970
PB 211,323 August 1972
PB 202,570 March 1971
PB 211,161 March 1972
PB 217,062 August 1968
PB 198,227 November 1970

PART I. DESIGN AND OPERATION

Incineration - A Perspective

The material in this chapter was based on a 1973 EPA publication, PB 223,626 prepared by A.F. Sarofin, G.C. Williams, J.B. Howard and J.E.L. Rogers of Massachusetts Institute of Technology.

Solid wastes are generated in five major areas: industrial, agricultural, commercial, household and institutional. The health and safety of the community in highly populated urban areas, where the disposal of the commercial, institutional and household waste is of major concern, have necessitated the development of municipal disposal systems. The general aversion to spending money on what is discarded as worthless and the apparent out of sight, out of mind philosophy of urban residents have made it evident to local governments that the first criteria for any waste disposal program would be low cost and ease of operation.

Rising sanitary and environmental standards and increased public interest in the quality of the environment have added to these criteria of acceptability. Public concern over air and water pollution problems and the depletion of natural resources, among other considerations, has handed technology a serious challenge to improve the methods of solid waste disposal.

Three hundred billion pounds of solid wastes are collected by municipalities throughout the United States annually and the problem of how to dispose of them in an acceptable way has been receiving a surge of attention over the past few years. Considerable impetus and financial backing for this long-needed attack on the mounting problem of solid waste were provided by the Solid Waste Disposal Act, PL 89-272, of October 1965.

Consequently, the solid waste management area has attracted the attention of a number of individuals and companies and several ways to deal with the old problem have been suggested. These have included high-temperature incineration, compaction followed by either landfill or offshore dumping, shredding followed by either burning or composting, as well as numerous ways of sorting out and classifying the reusable portions of solid wastes.

Traditionally, there have been three main methods of disposing of solid waste: open dumping, landfill and incineration, with dumping and landfill being the more widely practiced. The disposal of refuse by open dumping is an old concept. In the Middle Ages the streets of European cities were used as a dumping ground, and even in the Golden Age of ancient Greece the narrow crooked streets of Athens were heaped with refuse. The concept of the modern sanitary landfill dates back to Biblical times, when wastes were disposed of by burial.

Furnaces designed to burn refuse were not built until the end of the nineteenth century, with Nottingham, England, being credited as the birthplace of the first municipal incinerator in 1874.

Of the refuse collected in the United States in 1967, 86% was disposed of at landfill sites, 8% was burned in municipal incinerators and 6% was disposed of in sanitary landfill operations. Much of the refuse disposed of in landfill sites other than by sanitary landfill is ultimately burned under poorly controlled open-burning conditions.

The quantity of solid waste disposed of by each method has been traditionally determined primarily on the basis of cost. Concern, however, over the quality of the environment has generated a willingness to spend higher sums of money on solid waste disposal, in particular on methods that involve recycling reusable portions of solid waste. Act PL 89-272 was instrumental in promoting recycling, as it specifically outlined that attention be given to the conservation of natural resources by reducing waste and unsalvageable material.

It is expected that the portion of solid waste that is recycled will increase over the next decade or so, as it becomes apparent that the economic rents associated with environmental decay and the depletion of natural resources are deemed to be greater than the costs of other disposal methods. Recycling of solid wastes is not a new idea, but the low resale value of the salvageable items (rags, paper, glass, rubber and tin cans) made these operations economically unsuccessful except as sidelines for other methods of disposal.

Conceptually, the recycling of all the reusable portions of solid waste is the most attractive method of disposal but attainment of this halcyon position will take the time required for research effort and public education and acceptance. It is unlikely that recycling will be considered competitive with conventional methods of disposal for at least a decade, and until it is practiced on a wide basis these methods will need to be used and improved to meet the challenge of legislated environmental standards.

Even when recycling has become a major method of disposal, there will be significant quantities of waste that will still be unseparable, unusable, or which will require costly pretreatment for recycling, and these will have to be disposed of by the other conventional methods. For example, of three tons of refuse collected from three locations in San Francisco, only between 29.2 and 32.2% was recyclable.

Of the conventional methods presently employed, sanitary landfill has become more popular over the past years in response to public indignation over poor landfill and dumping practices which produce unsightly, odorous areas often infested with rats and vermin. The method of disposal is not without problems, specifically, accidental explosions, pollution of well water, and the increasing cost of the earth cover required.

The anaerobic decomposition of the organic portions of the waste produce, among other gases, methane, whose concentration may build up in a confined enclosure to produce an explosive mixture. Dangerous explosions have occurred or persons have been seriously burned as a result of accidental ignition of methane gas which had accumulated near old landfills or refuse dumps. These problems detract from the most attractive aspect of sanitary landfill, which is the eventual use of previously unusable land after it has been filled in.

In many major metropolitan areas, available landfill and dumping sites are becoming scarce and this dilemma has prompted the shipping of metropolitan waste to outlying communities. This practice, in turn, has come under sharp criticism from communities that understandably resent having to act as a dumping ground for others' problems.

The pressure for more landfill sites will continue to mount. Many metropolitan areas are experiencing rapid growth and the per capita daily waste in the U.S., a function of the rising standard of living, is steadily increasing. Both of these facts suggest that disposal practices requiring large land acreages and costly earth cover may become uneconomical and impractical. The only practical alternative in these areas may be incineration.

Incineration at this time is inefficiently practiced, as indicated by an analysis of a national solid wastes survey reported in 1970, which revealed that of the incinerators in the United States, 75% were inadequate, as far as air pollution control and effective residue burnout were concerned. Such statistics are not surprising, since 70 to 75% of the incinerators in the United States were built before 1960, with some dating back as far as the 1940s. Poor incinerator operation has caused the public to regard an incinerator as a bad neighbor and, although emissions from the burning of all solid refuse contribute less than 10% of the 11 million tons of particulates emitted to the atmosphere annually, they are the frequent cause of local pollution complaints by citizens.

In addition to the air pollution problem, the poor control of municipal units, which results in incomplete combustion, yields neither full volume reduction nor an inert residue. The residue eventually must be landfilled but, if it has been rendered completely inert, none of the landfill problems mentioned previously should occur, and the life of landfill sites could be dramatically extended.

Introduction to Incineration

The material in this chapter is based on an EPA publication PB-200 514, June 1971, prepared by J.R. Stear. The following is a discussion of incineration of municipal refuse in incinerators that are owned and/or operated either by governmental or nongovernmental groups. Such incinerators are large when compared to most nonmunicipal such as domestic, industrial and special purpose incinerators.

A study of incinerators in operation and under construction shows that in the past , the United States concentrated mainly on volume reduction. At the same time, European countries were not only concerned with volume reduction, but also the use of refuse as a fuel for steam and power generation. Air pollution control devices for removal of particulate matter are the concern of every country. Electrostatic precipitators, used extensively in Europe, seem to offer one of the better solutions for highly efficient emission control on municipal incinerators in this country.

Alternative refuse disposal methods are mentioned briefly, along with advantages and disadvantages of each method compared to incineration. Such a comparison should be of assistance in the evaluation of the method of incineration for a given locality.

INCINERATION AND OTHER REFUSE DISPOSAL METHODS

Many factors are involved in the evaluation of municipal incineration as a basic refuse disposal method. Some of the factors are dependent on the city in which an incinerator is used, and the remaining factors are an inherent part of the basic incineration process. Municipal incineration is generally considered to be an economical, nuisance-free, sanitary refuse disposal method. The determination of whether incineration is the best solution to a city's refuse problem is perhaps best reached after a careful consideration of its advantages and disadvantages, followed by an investigation of the advantages and disadvantages of other refuse disposal methods.

Advantages of Municipal Incineration

There are ten basic advantages expressed by various authors in the refuse disposal field:

(1) Municipal incinerators require relatively small plots of land as compared to sanitary landfills.
(2) Incinerators can usually be located in industrial areas near the center of the service area and near collection routes.
(3) The incinerator operation is not interrupted by inclement weather.
(4) The operating time of a municipal incinerator can range up to 24 hours a day to accommodate the variations in refuse generation.
(5) The residue from an incinerator is generally stable and nearly inorganic.
(6) An incinerator can be designed to be inconspicuous to allow it to be located in or near residential areas.
(7) Incinerators located near sewage treatment plants can complement the plants by burning malodorous gases and drying and burning sludge.
(8) Incinerators can be less expensive than sanitary landfills.
(9) Incinerators can burn practically any kind of refuse.
(10) Income can be realized if there is a market for steam, hot water, electricity, salvage metals, residue, and incineration service to industry.

Other advantages undoubtedly have been realized in municipal incineration.

Disadvantages of Municipal Incineration

Little has been written on the disadvantages of municipal incineration, which indicates the disadvantages may be few.

Perhaps the biggest disadvantage to incineration is the high initial cost of an incinerator facility. Even though long-term landfill cost has been reported to be as much or more than incineration, the initial cost of a plant can be a substantial burden to a municipality.

Municipal incinerators emit pollutants into the atmosphere. Adherence to air pollution control requirements are becoming mandatory nearly everywhere in the world. Efficient means to control air pollutants released to the atmosphere are available, but are relatively expensive.

Operating costs are relatively high. Although the number of employees required to run an incineration plant may be smaller than it is for other methods of disposal, the wages for the skilled employees who operate, maintain, and repair an incinerator are higher, for instance, than for men who work on a landfill.

Maintenance and repair costs may be high because of the high temperatures necessary for the burning and the dirty and damaging nature of the refuse and residue. Equipment and machinery are frequently damaged by wires; tramp metals; and fusible, abrasive, and explosive objects in the refuse. The combination of large capital investment, higher labor costs, and costly maintenance and repairs can produce a cost per ton for refuse disposal greater than for other acceptable methods.

It is sometimes difficult to get a site for an incineration plant because refuse disposal operations in any form are offensive to many people. Moreover, truck traffic to and from the plant may be considered a hazard and a nuisance, particularly in residential neighborhoods. Incineration is not a complete disposal method. Ash and other residue from the burning process, including flyash, must be disposed of by other means.

Other Disposal Methods

A brief review of dumping, open burning, sanitary landfill, composting, burial at sea, disposal in sewer, and hog-feeding will provide some perspective in the evaluation of municipal incineration. Some of these methods could provide for the complete disposal of municipal refuse while others would provide disposal of only a fraction of the refuse.

Dumping: One of the first refuse disposal methods used was open dumping of refuse on land. This method is still widely used and is obviously very inexpensive, but extremely objectionable and offensive in and near populated areas. Such a site is a breeding ground for insects and rodents; odors can pollute the air for considerable distances from the dump. Long hauling distances are required to reach areas where few people live, and even then objections still arise.

Open Burning: Open burning has all of the disadvantages of dumping without burning, except that the volume of the refuse is reduced and less land is required. The insect and rodent problem is reduced somewhat; however, the burning produces extensive amounts of smoke and increased odors.

Sanitary Landfill: A sanitary landfill can be an economical and nuisance-free method of refuse disposal. A sanitary landfill involves burying refuse in a sanitary manner. Unsightly refuse is hidden from view by a layer of earth that controls the insect and rodent problem found in open dumps. Sufficient land at a reasonable price and in close proximity to the refuse collection area must be available to make a landfill successful. Public acceptance is sometimes difficult to achieve because a sanitary landfill is often associated with open burning and dumps or landfill operations that are not sanitary. Sanitary landfills can be an excellent method for filling depressions, canyons, tidal areas, swamps, and marshes.

Some of the problems associated with landfills are adverse weather conditions, dust, odors, water pollution due to leaching, settlement of the landfill, and formation of explosive gases in the decomposing section of the fill. A completed landfill can be successfully used for recreational purposes, parking areas, and construction provided some important precautions are taken.

Composting: Composting is the biochemical degradation of organic materials to a sanitary, nuisance-free, humus-like material. Composting has potential as a refuse disposal method, but has not been used extensively enough to permit an evaluation of its operational effectiveness. Advantages of composting based on pilot plants operating in Europe and the United States during the 1950s are as follows:

(1) A composting plant produces a usable end product that may be sold, thus either paying for, or at least reducing, costs.

(2) Composting can be used to dispose of such industrial wastes as those from meat packing plants, paper mills, saw mills, tanneries, stockyards and canneries. Dewatered sewage solids, especially if they are mixed with ground refuse, may be disposed, and cans and bottles that have no salvage value can be economically ground with the remaining refuse. When large grinders are used (plants capable of processing more than 100 tons a day), even large, bulky objects may be handled. A municipal refuse composting plant can dispose of all these wastes.

(3) Normally, composting offers favorable conditions for salvage of rags, glass, cardboard, paper, cans and metals.

(4) A well-located refuse composting plant may reduce the cost of hauling refuse to the point of disposal.

(5) Flexibility of operation permits a 100 to 200% overload in design capacity for several days by increasing the time the receiving bins and grinders operate.

(6) Weather does not affect an enclosed composting plant, although heavy rain adversely affects most kinds of outdoor composting.

Disadvantages to composting include:

(1) Capital and operating costs apparently are relatively high.

(2) Marketability of composted refuse has not yet been proved and seasonal use of the end product may require special marketing procedures or outdoor storage.

(3) Trained personnel to operate composting plants are not readily available.

(4) Refuse that damages grinders, such as tires, pipes, heavy stones, and mattresses, must be removed and disposed of separately.

(5) If cans and bottles have no local salvage value, they must either be removed and disposed of separately or ground with the organic matter, thus somewhat reducing the quality of the finished compost.

(6) Site procurement for a composting plant is difficult because any type of refuse disposal facility is considered a nuisance in most neighborhoods.

(7) Odors can become a problem during periods when a compost plant is not functioning properly.

Dumping at Sea: Dumping of raw refuse at sea was practiced rather widely prior to 1953 by New York City and other communities. After many communities were recipients of floating refuse, however, a United States Supreme Court decision in 1953 prohibited dumping of the raw refuse at sea. Some consideration has been given in recent years to incineration aboard ships at sea, with residue disposal in the sea. Unknown effects of the residue on sea life and the possibility of floating debris make this method questionable.

Disposal in Sewer: This method, which is practiced ever more widely in private homes, consists of grinding the refuse on a municipal scale, metal and glass excluded, and then disposing the ground refuse into a city's sewage system. This method is generally not considered to be very feasible. Most sewage systems are already overburdened. Additional equipment and experimentation that would have to be performed to put such a system into operation would be quite costly.

Unit Trains: The use of a special train carrying only refuse as cargo (unit train) has potential as a city's refuse disposal system. Such a train would carry the refuse to an unpopulated area such as a desert, swamp, or mountainous area where it would be dumped. This type of system may still meet with some of the serious problems encountered with local open dumps.

Swine Feeding: Feeding of cooked food wastes to swine can accommodate only a fraction of the total refuse produced by a municipality. Opposition to hog-feeding has grown, as evidenced by increased zoning restrictions and the elimination of this garbage disposal method in many areas. All garbage must be cooked before it is fed to swine to help prevent vesicular exanthema, hog cholera, and enteritis diseases. A high correlation exists between trichinosis in humans and the feeding of raw garbage to swine.

REFUSE INCINERATOR TYPES

Numerous methods of classification of municipal incinerators are available. One classification system is based on refuse disposal capacity. Another classification system is based on use of by-products. The discussion of incinerators herein will be according to the method used to feed refuse to the furnace.

Continuous Feed Incinerators

The continuous feed incinerator is used almost exclusively for municipal incineration. Such an incinerator moves the refuse automatically from a hopper through the furnace on a grate (stoker). Refuse burns on the stoker as it passes through the furnace. Refuse is ignited while on the feeder grate, then tumbles off the feeder grate onto the burner grate where rapid combustion takes place. Some incinerators have more than one burner grate, each of which is successively lower so that tumbling of the refuse from grate to grate further exposes unburned portions of the refuse to the combustion process, thereby assuring more complete combustion.

Traveling Grate Incinerator: The traveling grate continuous-feed incinerator consists of a continuously moving feeder grate and one or more burner grates. The feeder grate is located directly under a charging hopper from which refuse falls onto the grate. The refuse can be partially dried while it is on the feeder grate.

FIGURE 2.1: MUNICIPAL INCINERATOR IN THE CITY OF DUSSELDORF, GERMANY

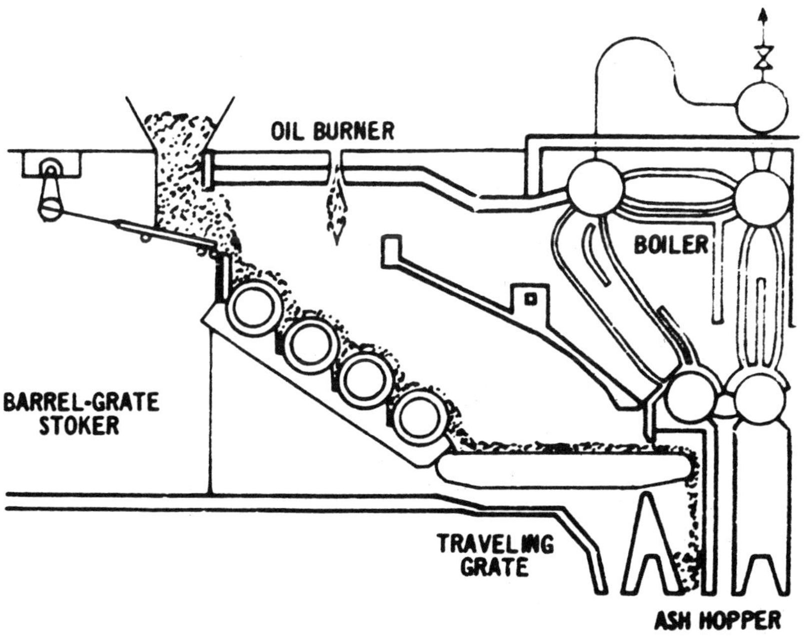

Reciprocating Grate Incinerator: The reciprocating-grate incinerator moves refuse through the furnace from the hopper while the grate is actually stationary, except for alternating reciprocating movements of component stoker bars. The action of the stoker bars turns the refuse over and then tumbles it forward to the next successive stoker bar. Burning rate is adjusted by controlling the speed of the stoker bars.

Rotary Kiln Incinerator: Refuse is dried and charged into the rotary kiln incinerator in the same manner as for the traveling grate incinerator. The difference in the two incinerators is in the burner grate. The refuse is dumped from the feeder grate into a rotary kiln that provides constant tumbling of the burning refuse. The kiln is continuously charged and provides continuous residue removal.

Barrel Grate Incinerator: The barrel grate incinerator is relatively new in design. Refuse is burned as it is moved by a series of rotating barrels. One such incinerator is in operation in Dusseldorf, Germany (Figure 2.1).

Batch Feed Incinerator

The batch feed incinerator is as its name implies, noncontinuous. Refuse is charged to the furnace through the furnace roof at periodic intervals to allow the previous batch to be almost completely burned, when a new batch is introduced. The residue is normally removed from the furnace in the batch method, but at a frequency much lower than the batch charging. Some installations have provisions for automatic removal of residue and incombustible components. Automatic agitators often provide constant overturning and mixing of the burning refuse to allow a minimum of hand stoking.

Cell incinerators, used more extensively in Great Britain and Europe than in other countries, can be considered batch incinerators. Normally, the furnace is made up of from 2 to 6 cells. Each cell receives a premeasured charge of refuse through the charging gate. This refuse is then dropped onto the horizontal grate where it can be stoked by hand while burning. The grate is mechanically controlled for residue removal.

Air for combustion is heated as it passes over the hot residue that has been removed from the horizontal grate. All cells are open to each other and adjacent to the last cell in the direction of the combustion gas stream is a combustion chamber where the combustion of the gases is completed.

Ram Feed Incinerator

The ram feed incinerator uses a ram to move the refuse from a charging hopper to the burner grate. The burner grate then moves the burning refuse continuously through the furnace. Residue and noncombustibles are removed continuously at the end of the burner grate.

Metal Conical Incinerator

Although of the batch type, conical incinerators are discussed separately because of their distinctive design and undesirability for municipal refuse incineration. Metal conical burners are similar in shape to an Indian tepee. A survey of 15 burners in six states found that the size of the burners range from 10 feet in diameter by 12 feet high to 90 feet in diameter by 97 feet high. The base of the conical incinerator is usually secured to a concrete ring foundation. Walls are usually made of 16-gauge steel. Some also have an inner lining of steel. The dome is usually fitted with a mesh wire for collection of large particles of flyash. Many draft doors are located at the base of the burner. Most conical burners are equipped with forced-draft blowers. Charging may be done by conveyor belt, bulldozer, or elevated truck chute.

Waste Heat Recovery Incinerators

The calorific value of refuse in the United States averages about 5,000 Btu/lb. Of this amount, about 45% is usually released as waste heat to the atmosphere through the stack. Only a very few incinerators in the United States are designed to recover waste heat. In Europe, however, most of the large modern municipal incinerators built since World War II are designed to recover waste heat. Waste heat is recovered by use of low pressure boilers, high pressure boilers, and, most recently, water walls. Four elements that have made waste heat recovery practical, efficient, and economical in Europe are:

(1) Development of more effective and efficient incinerators to handle refuse that is difficult to burn and low in heat value.

(2) Development of more effective heat recovery systems.

(3) Recognition of the considerable aid given to the alleviation of air pollution from the incineration of refuse. Approximately 50% of the particulate matter is removed by the average waste heat system. Furnace gases of from 1800°F are cooled to the required 400° to 500°F and gas volumes are reduced by at least one-half.

(4) Continued unavailability of economically competitive fuels.

Recovered waste heat can be used to produce steam for heating or for the production of electricity and hot water for heating, personnel services, and process requirements. Waste heat can be used to dry sewage sludge, which can then be sold as a fertilizer, and at coastal sites it can be used to desalt salt water from the ocean to supply communities with potable water. Waste heat recovery incinerators are generally of the continuous type that burn large amounts of refuse, the waste heat output of which is fairly constant and dependable.

Low Pressure Boilers: Hot water can be generated in low pressure boilers that are heated by hot gases that pass from the furnace to the boilers and then back to the stack. In refuse plants that use the hot water for internal heating and service requirements, only a small portion of waste heat is recovered. Because there is always a demand for hot water, external demands of municipal buildings and factories should not be overlooked.

High Pressure Boilers: Early designs of waste heat recovery incinerators placed boilers in the combustion chamber with direct exposure to the burning refuse. The absorption of heat by the boiler, together with lower calorific value of refuse at the time, lowered furnace temperatures and thus the effectiveness of the combustion. To alleviate this problem, boilers built directly above the burning refuse are shielded to prevent excessive cooling of the furnace by radiation.

Many refractory incinerators that pass combustion gases through a series of boilers have been designed (Figure 2.2). Not only is waste heat recovered, but the volume of the combustion gases needing cleaning is reduced considerably. Boilers may be used to reduce flue gas temperature to within the range (482° to 572°F) that such high grade dust collectors as electrostatic precipitators cannot be subjected.

FIGURE 2.2: CROSS SECTION OF A LARGE EUROPEAN INCINERATOR, SHOWING PATH OF FURNACE GASES THROUGH HEAT RECOVERY BOILER

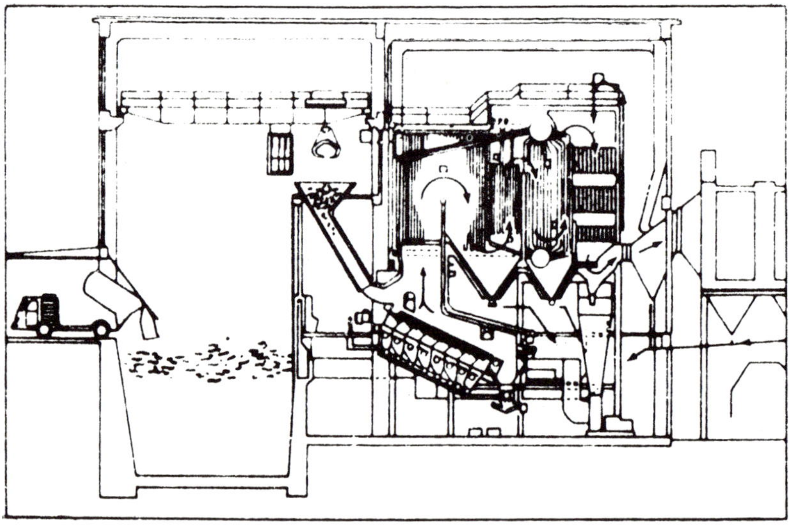

Water Wall Furnace: Water walls are used in furnaces in various European incinerators, a practice that is not new. The only operational water wall incinerator in the United States is located at the Norfolk, Virginia, Navy Base.

Water walls are constructed of interconnected steel tubes welded together to form an integral wall. Circulating water is converted to steam almost entirely by radiation supplemented by some convection. Obviously, many factors are involved in the amount of heat that is transferred from the furnace chamber to the water walls. If too much water wall area is installed, the furnace may operate at temperatures below deodorizing temperatures, resulting in an undesirable situation. As with boilers, the cooling of the furnace by water walls means a lower requirement for the quantity of excess air, resulting in less flue gas requiring air pollution control treatment. A cross section of the Norfolk refuse incinerator is shown in Figure 2.3. The specifications for each boiler-furnace unit are given in Table 1.

FIGURE 2.3: WATER WALL INCINERATOR AT NORFOLK, VA.

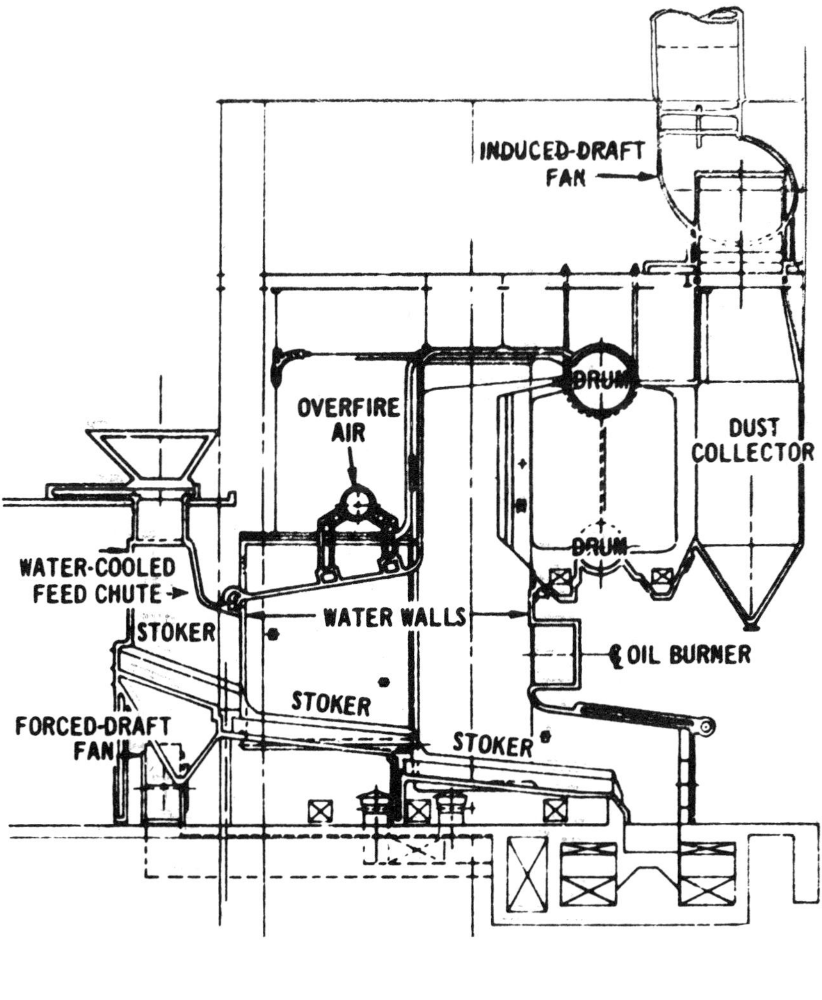

TABLE 1: SPECIFICATIONS FOR EACH BOILER-FURNACE UNIT IN WATER WALL INCINERATOR, NAVY BASE, NORFOLK, VA.

Mixed refuse capacity, tons/day	180
Heat content, Btu/lb as-fired	5,000
Moisture, percent	25
Noncombustible material, percent	12.5
Steam production	
With refuse at 5,000 Btu/lb, lb/hr	50,000
With drier refuse or with refuse plus oil, lb/hr	60,000
With oil only, lb/hr	50,000
Design stoker loading	
lb refuse/ft^2-hr of effective grate surface	65
Heat release	
Btu/hr-ft^2 effective grate surface	325,000
Btu/ft^3 primary furnace volume maximum	25,000
Minimum gas temperature leaving primary furnace at 50% of rated load °F	1,400
Steam pressure, psig	275
Steam quality	Saturated
Feedwater temperature, °F	228
Exit gas temperature from boiler at design refuse capacity, °F	580

Salt Water Distillation: Waste heat can be used effectively to desalinate water. Its use in providing a future water supply for coastal areas should not be underestimated. Experience gained from the Oceanside Refuse Disposal Plant shows that waste heat from the incineration of refuse from half a million people could supply one-fourth to one-third of their domestic water requirements.

The basic principle is that steam is generated from waste heat. The steam is used to convert the salt water, pumped from a nearby source, to steam. The salt water steam is then condensed to fresh water on tubes that are cooled with unheated salt water. A flow diagram of the Oceanside Refuse Disposal Plant is given in Figure 2.4.

Sewage Sludge Disposal: Refuse incinerators can be located adjacent to sewage treatment plants where waste heat can be used to dry sludge or where dewatered sludge can be mixed with the refuse and burned. Dried sludge can be sold as a fertilizer. Effluent water can be used for cooling the incinerator furnace walls and for gas scrubbing. Raw sludge is dewatered by a vacuum filter.

The resulting sludge cake contains 65 to 75% moisture. The moisture content can be further reduced by storing the sludge for several days. Heating values, ash content, and percentage of volatile matter for a typical sludge cake are given in Table 2.

FIGURE 2.4: FLOW DIAGRAM TRACING WASTE HEAT AT WORK GENERATING STEAM FOR POWER AND DESALTING

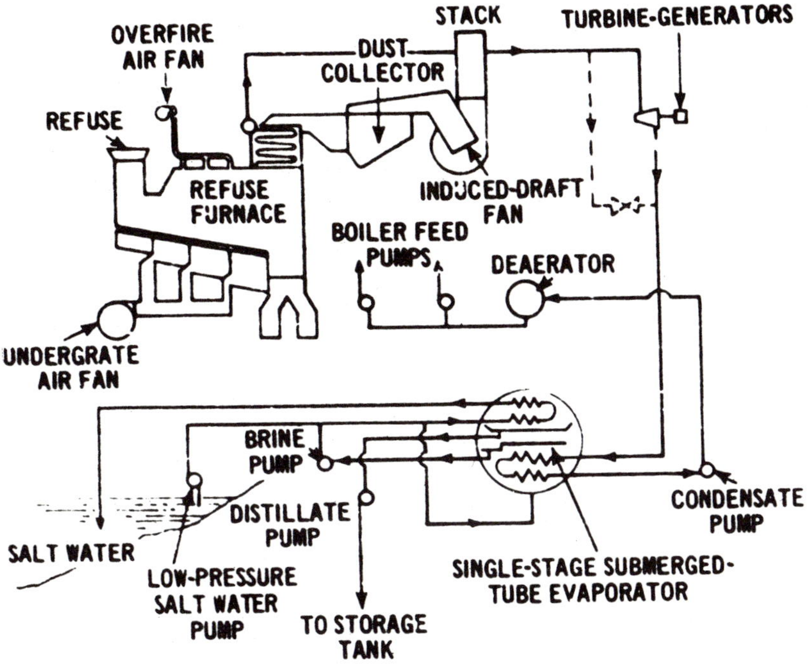

One such sludge-burning incinerator is located in a suburb of Philadelphia, Pa. Oscillating conveyors are used for mixing the refuse and sludge. A performance test conducted on this incinerator showed that for 381.4 tons of refuse burned, 32.0 tons of sludge were burned. Combustible material that was not destroyed amounted to 4.36%.

TABLE 2: AVERAGE SEWAGE SLUDGE CHARACTERISTICS

Moisture, % of sludge cake	
Range	65 - 70
Average for design	70
Volatile matter including chemicals, % of dry solids	
Range	50 - 85
Average for design	70
Ash content including chemicals and combustibles, % of dry solids	
Range	50 - 15
Average for design	30

(continued)

Heating values, Btu/lb	
Dry solids, range	5,600 - 10,000
Combustible design average	11,500
Combustible in ash, % of ash	
Maximum allowable	4

OPERATION OF MUNICIPAL INCINERATORS

Operating Temperatures and Their Measurement

Actual flame temperature inside of the combustion chambers of municipal incinerators is approximately 2400°F. The furnace temperature, commonly referred to is the temperature of combustion gases exiting the combustion chamber. Except in water walled furnaces, furnace temperature is usually controlled to the range of 1800° to 2000°F. If truck loads of refuse with unusually high or low calorific value are not mixed with the normal refuse by the crane operator, this approximate range of temperatures can be several hundred degrees too low or too high.

A load of sawdust in a charge, for example, could increase the normal operating temperature of 1800° to over 2000°F in approximately 15 seconds. While this temperature fluctuation may be most prevalent in batch-feed furnaces, it also exists in continuous-feed furnaces.

When furnace temperatures are referred to, the precise location at which the temperature is taken in the furnace should be noted. As already discussed, furnace temperatures are usually controlled by using excess air, that is, air in excess of that needed to completely burn the combustible portion of the refuse. The effect of excess air on furnace temperatures for various moisture contents is shown in Figure 2.5.

From the combustion chamber, the gases enter either a waste-heat boiler area, gas-cleaning devices, or cooling towers. At this point, the gas has cooled usually to within the range from 1400° to 1800°F. In incinerators equipped with waste-heat boilers or cooling towers, the temperature of the gases is from 500° to 700°F after passing through the device. On leaving the gas-cleaning devices, the temperature of gases that have not previously been cooled by waste-heat recovery or by passage through cooling towers is usually less than 1000°F.

Incinerator temperatures are usually measured with either electrical or filled-bulk sensing devices. Among the electrical types applicable to incinerators are the thermocouple, thermopile, radiation pyrometer, and thermistor. The thermocouple and the thermopile, which is made of thermocouples arranged to produce a higher electrical output, are perhaps the most widely used temperature-measuring devices used in incinerators. They can measure temperatures of from 2000° to 2300°F. Radiation pyrometers can withstand higher temperatures than thermocouples, and are, therefore, normally used for measuring actual flame temperature in the combustion chamber. Pyrometers normally have effective temperature-sensing ranges of from 1000° to 4000°F. Thermistors of platinum wire are applicable for temperatures ranging from 400° to 1000°F.

FIGURE 2.5: RELATIONSHIP OF MOISTURE, EXCESS AIR, AND FURNACE TEMPERATURE

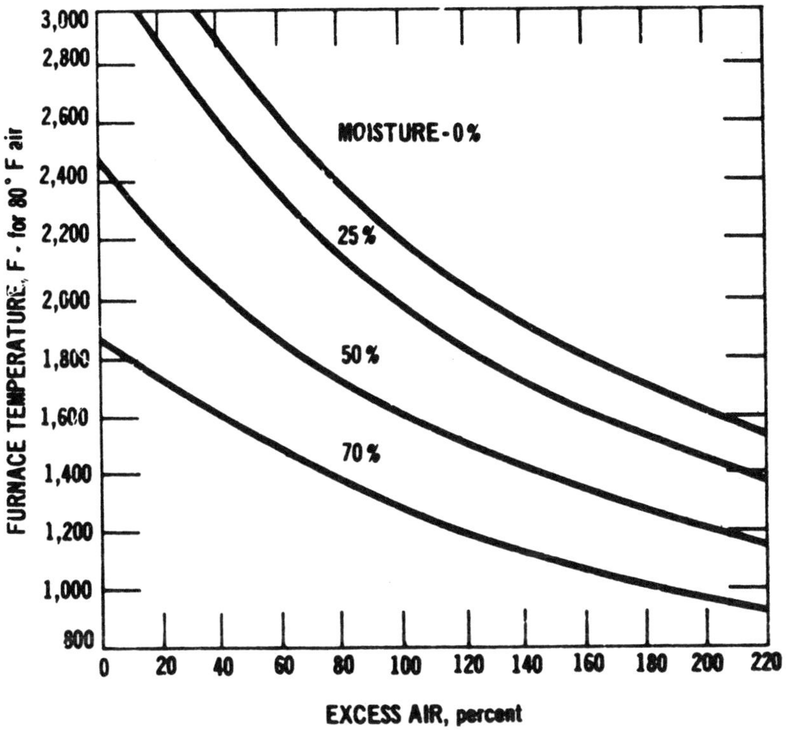

Filled-bulb temperature-sensing devices using the expansion properties of a liquid, vapor, or gas are not used as widely as electrical-sensing devices in incinerators. Compared to the electrical devices, they are quite large and bulky and require extensive shielding to protect them from corrosion, erosion, and deposition. Filled-bulb devices can measure temperatures up to 1200°F.

Thermocouples are usually recommended for temperature measurements at the combustion chamber exits and in the flues just prior to waste-heat boilers or cooling towers. Protecting wells must be provided for the thermocouple. The wells are often coated with silicon carbide to protect them from corrosion from slag. Thermistors and filled-tube devices can be used between either waste-heat boilers or gas-cleaning devices and stacks.

Pyrometers are the only practical devices for measurement of actual flame temperature. Because pyrometers are expensive and the relationship among fire temperature, refractory life, and combustion is not well understood, the fired temperatures are not usually measured.

Operating Pressures and Draft Requirements

The air supply and combustion gases in incinerators are controlled by forced, natural and induced drafts. A forced draft is created by the pressure difference generated by a mechanical device that supplies air at a pressure greater than atmospheric pressure. Forced drafts are used to supply primary combustion air to the drier and burner grates of incinerators. Natural draft results from the pressure difference created by a stack and is a function of the stack height and temperature difference between the flue gases and the ambient air. Natural drafts are normally neither sufficient nor consistent enough to remove combustion gases from large incinerators, particularly those that are equipped with gas-cleaning devices.

Induced drafts are the result of the pressure difference created by a mechanical device located between the furnace and the top of the stack. Short stacks require induced drafts. Air pollution control devices have various ranges of pressure drops that must be taken care of by induced-draft fans. The draft requirements for air pollution control equipment are shown in Table 4, page 107 in inches of water column.

Operational pressures differ from one incinerator to another in such a manner that average pressures would have little meaning. Draft gauges are usually located to measure primary and secondary air pressure, furnace pressure, induced-draft suction pressure, and pressures at the inlet and outlet of various air pollution control devices.

Management

Schedules: Municipal incinerators are routinely operated for periods of from 8 to 24 hours a day and for 5, 6 or 7 days per week. A survey in which 154 incinerators in the United States reported operating schedules is summarized in Table 3.

TABLE 3: SUMMARY OF OPERATING SCHEDULES OF 154 INCINERATORS

Operating Period, hour/day	Number of Plants	Percent of Total
8	55	36
9	2	1
10	2	1
12	1	1
16	9	6
18	1	1

(continued)

TABLE 3: (continued)

Operating Period, hour/day	Number of Plants	Percent of Total
20	2	1
24	82	53

The current trend in operating schedules is toward 24-hour-a-day operation. Table 4 shows operating schedules of two groups of incinerators, based on year of construction. The table indicates that this trend perhaps began around the year 1964.

TABLE 4: DAILY OPERATION OF 154 MUNICIPAL INCINERATORS

Year of Incinerator Construction	Number of Plants Reporting	8 Hours #	8 Hours %	9-20 Hours #	9-20 Hours %	24 Hours #	24 Hours %
1945 through 1963	117	48	41	13	11	56	48
1964 through 1966	37	7	19	4	11	26	70

Personnel: The number and types of personnel vary with the operating schedule, design,and degree of automation of the incineration plant. One versatile maintenance man can sometimes perform tasks normally requiring knowledge in two or more diverse fields. Operating personnel of some modern incinerator plants are discussed in the following paragraphs.

Rockville, Maryland Incinerator: The Montgomery County, Maryland (Rockville) incinerator has a capacity of 1,050 tons/day. It operates 24 hours a day, 6 days a week. The 44 operating personnel include:

1 plant supervisor
4 foremen
9 equipment operators (bulldozer and crane operators)
3 truck drivers
8 furnace stokers
12 laborers
1 clerk
1 weighmaster
1 electrician
1 plumber
1 mason
1 general maintenance man
1 janitor

Detroit, Michigan Incinerator: Northwest Incinerator Plant at Detroit has two furnaces, each of which is capable of burning 425 tons of refuse every 24 hours. The plant operates three shifts a day, 5 days a week. The 51 employees include:

Operating crew per shift
1 incinerator foreman
3 electric crane operators
6 incinerator firemen
1 charging floor man
1 ash tunnel man
3 semitruck drivers

Additional personnel on day shift
1 scaleman
1 tipping floor man
1 mechanical tradesman
3 janitors

Milwaukee, Wisconsin Incinerator: The Lincoln Avenue Plant in Milwaukee, Wisconsin, has a rated capacity of 300 tons/24 hours. The plant is operated on a 5-day-a-week schedule. Machinery operator service is provided on a 7-day-a-week basis. Table 5 shows how personnel at the plant are used.

TABLE 5: LINCOLN AVENUE PLANT OPERATING PERSONNEL

	----- Number of Employees -----		
Job Title	**1st Shift**	**2nd Shift**	**3rd Shift**
Operating engineer II	1	1	1
Craneman	1	1	1
Furnaceman	3	3	3
Disposal division laborer	4	3	3
Machinery operator	1	1	1
Collection division laborer	2		
Electrical mechanic	1		
Maintenance mechanic	1		
Truck driver	1	1	
Mechanic helper		1	1
Total	15	11	10

Engineer-in-charge — In charge of all disposal department activities. Asst. Engineer-in-charge — Assistant to above, and directly responsible for all maintenance and repair. Firebrick Mason — Maintenance and repair of all refractories. Electrical Mechanic — Maintenance and repair of all electrical equipment. Operating Engineer II — In charge of one plant shift. Maintenance Mechanic Foreman — In charge of all maintenance and repair activities. Boiler Repairman — Maintenance and repair of waste-heat boilers. Craneman — Operation of overhead traveling cranes. Machinery Operator — Operation of waste-heat boilers, feed water pumps, oil burners, etc. Maintenace Mechanic I — Maintenance and repair of all mechanical equipment. Furnaceman — Operation of furnaces, dampers, temperature controls, draft fans, charging gates, etc. Truck Driver — Operation of incinerator ash removal truck. Mechanic Helper — Assistant to maintenance mechanic, electrical mechanic, boiler repairman, or firebrick mason.

Washington, D.C. Incinerator: Mount Olivet incinerator in Washington, D.C. has a 500-ton capacity. The incinerator consists of four 125-ton furnaces of the mutual-assistance type with rocking grates. It has a peak capacity of 700 tons.

TABLE 6: MOUNT OLIVET INCINERATOR OPERATING PERSONNEL

Job Title	**8–4**	**4–12**	**12–8**
Plant foreman	1	1	1
Mechanic	1		

(continued)

TABLE 6: (continued)

Job Title	8-4	4-12	12-8
Overhead crane operator	2	2	2
Incinerator fireman	4	4	4
Equipment lubricator	1		
Weighmaster	1	1	
Laborer (hopper)	2	2	2
Laborer (ash tunnel)	2	2	2
Laborer (janitor)	1		
Laborer (watchman)	1	1	1
Refuse transfer operator (platform)	1		
Incinerator driver	2	2	
Total	19	15	12

The St. Quen incinerator in Paris, France, consists of four furnaces with a total annual capacity of 407,000 tons and recovers waste heat. Approximately 160 people are employed. The Stuttgart incinerator in Germany, with an annual capacity of 220,000 tons, has a staff of 55 employees.

Municipal incineration is usually organized under a municipal government but is occasionally a function of a county government or an autonomous incinerator authority operating under a governmental charter. Usually plant supervisors, and sometimes the maintenance men, are hired during the planning or construction phases so that they become intimately familiar with the plant.

Immediately prior to the opening of a plant at least one person for each job classification is hired and instructed on the operation of the plant. Actual performance tests are usually made by the manufacturer on the individual pieces of equipment.

Maintenance

Plant Maintenance: Neatness and proper maintenance of building and grounds are at least as important to the aesthetic quality of incinerators as they are to the aesthetic quality of other industrial buildings. Refuse disposal facilities are not generally held in esteem by any community, and a littered, unmaintained facility can provide justification for such sentiment.

Janitors are usually a part of the full time staff at most incinerators. Their duties are usually routine, but very important to the cleanliness of the plants since incinerators' environments can be quite dusty. Some plants use a central vacuum system with connecting outlets throughout the building. The central vacuum system can be connected so that it disposes of collected wastes directly into the residue-handling system.

Maintenance Facilities: Some incinerators are equipped with maintenance shops that are fully equipped with the power tools and equipment necessary to repair various types of incinerator equipment. Spare parts for cranes, stokers, fans, and motors, which are not always readily available as shelf items, are sometimes kept on hand. Most operational maintenance is performed by regular staff employees.

Preventive Maintenance: Serious problems can be prevented by frequent preventive maintenance. Weekend shutdowns provide an excellent opportunity to inspect for and detect future problem areas. Refractory maintenance, boiler care, slag removal, and grate maintenance are just a few of the important areas that should be serviced frequently. The maintenance facility of the plant is usually able to perform routine preventive maintenance.

Plant Safety: Plant safety is a continuing concern of incinerator designers and supervisors. Serious accidents, including loss of life, have occurred at some older plants. Incinerator hazards have been attacked on two fronts. One method of attack is to incorporate safety design features into the construction of the plant. Some of the safety features included when plants are constructed are:

(1) Brick and concrete building materials, automatic or manual sprinkler systems for storage pits and charging floors, and fire hose stations at strategic locations for fire protection.

(2) More space on stoking floors to avoid crowded conditions.

(3) More adequate lighting from large windows and artificial lights.

(4) Lunch rooms, locker rooms, and showers for more sanitary conditions for employees.

(5) Drinking fountains with salt tablet dispensers to reduce the number of cases of heat exhaustion during warm weather.

(6) Better pitching of floors with adequate drains to aid in cleanliness and prevent falls.

(7) Improved building ventilation. The use of outdoor suction intakes for forced-draft fans avoids the possibility of creating a vacuum on stoking floors that can cause blow-outs of flame through the stoking doors of ignition chambers with serious hazard to employees.

(8) Chimneys equipped with airplane lights, lightning rods, and safety ladders as standard equipment.

(9) Two-way radio systems for signaling between the charging and stoking floors to supplement other signaling devices and avoid confusion and mistakes in directions. For safe operation it is important that the charging doors be closed when stoking.

(10) Refuse storage pits with access ladders, and either forced fresh air inlets or mechanical exhaust systems with air ports located near the bottom of the pits. Decomposing garbage may deplete the oxygen necessary for life and cause oxygen deficiency in men entering such

pits. Pit drains are also standard practice to permit hosing of pits for thorough cleaning to avoid nuisance and free the pits of vermin.

(11) Elevators to carry heavy equipment from one floor to another and for employee convenience.

(12) Stationary vacuum pumps having port outlets at convenient locations for suction cleaning of floors, stairs, and flues.

The second method for attacking safety hazards is to practice and enforce such operational safety practices as:

(1) Keeping truck stops at the tipping edge of refuse storage pits in good repair, and prohibiting employees or others from standing on them when trucks are unloading.

(2) Using mechanical ventilation when refuse pits containing decomposing refuse must be entered by employees. Portable air blowers are used when provisions for mechanical ventilation have not been provided. (Safety belts should be worn when persons are descending long ladders.)

(3) With floor-charge furnaces, protecting the men charging the furnaces with safety belts or guards placed around the charging openings.

(4) Using an alarm in the crane housing of bucket-charged furnaces to warn employees, as necessary, of descending loads of refuse.

(5) Wearing of protective clothing, including safety shoes, heavy gloves, and goggles or face shields by stokers.

(6) Restricting entrance of flues for removal of flyash to times when the temperature is below 100°F. Respirators are worn for comfort.

(7) Providing first-aid kits and posting emergency instructions for employee information.

(8) Proper housekeeping.

Incineration Costs

COMPARING COSTS

There are innumerable ways in which cost data may be presented. The method that is best suited to the need at hand is usually chosen. Rather than try to interpret the needs of prospective readers, cost data is discussed and presented in this chapter to a large extent in the manner in which it appears in the published literature. Certain precautions are necessary for users of such data. For comparative purposes, costs should be reduced to a base year to adjust for inflationary trends. Cost differences of utilities, materials, and labor from one area to another or one country to another should be taken into consideration.

Design specifications such as control equipment emission criteria may have a very significant effect on the price of a plant. Sometimes expensive construction design is necessary because of unusual soil conditions or climate. The economy of building one large incinerator instead of two incinerators half the size can be significant even though transportation distance of the refuse may be increased and vulnerability of the overall operation to breakdowns is increased. These are just a few of many factors that make an accurate determination of costs, and comparisons of costs in one area of the country to those of another, difficult, if not impossible.

INITIAL PLANT CONSTRUCTION COSTS

Plant costs are most commonly given either as total cost or in cost per ton capacity per day. The total cost price can range anywhere from $52,000 for a simple 140-ton-per-week conical burner serving a town of approximately 15,000 to $30,600,000 for a 660,000-ton-per-year, steam-generating, continuous-feed incinerator. Capital costs for some modern incinerators are corrected

TABLE 1: INCINERATOR PLANT CAPITAL COSTS

Location	Furnace Types, number and size, tons/day	Supplementary Fuels	Water-Cooled Walls	Refractory Walls	Special Auxiliaries: Waste Heat Boilers	Special Auxiliaries: Electrostatic Precipitators	Special Auxiliaries: Steam Generating Plant	Special Auxiliaries: Electric Generating Plant	Special Auxiliaries: Oversized Weight Equivalent	Special Auxiliaries: Residue and Metals Salvaged	Chimneys per Plant	Capital Cost: Total Cost of Plant	Capital Cost: Unit Cost per Ton per Day	Remarks
Vienna	[a]Van Roll 3, 200	Waste oil	-	X	X	X	X	X	-	X	1	$ 9,600,000	$16,000	Difficult foundations
	[a]Martin, 2, 660	Pulverized coal	X	-	X	X	X	X	X	X	1			
Munich	[a]Martin, 1, 1,060	Pulverized coal	X	-	X	X	X	X	X	X	1	21,750,000	9,100	Costs are approximate
Dusseldorf	[a]Drum, 4, 250	-	X	-	X	X	X	X	X	X	1	7,500,000	7,500	Cost includes building for total of 6 furnace units
Rotterdam	[a]Martin, 4, 385	-	X	-	X	X	X	X	X	X	2	9,250,000	6,000	Steam for on-site only. electric power generator: 10% for inc. balance sold
Paris (Issy-les-Moulineaux)	[a]Martin, 4, 450	-	X	-	X	X	X	X	-	X	2	20,000,000	11,100	Highly sophisticated
Lousanne	[a]Van Roll, 2, 200	-	-	X	X	X	X	X	-	X	1	4,000,000	10,000	Handsome plant in midst of apartment area
Montreal	[a]Martin or Van Roll, 4, 300	-	X	-	X	X	X	-	X	X	2	12,000,000	10,000	Bids taken Nov. 2, 1965
N.Y.C. (6 modern plants)	[a]Traveling grate, 4, 250	-	-	X	-	-	-	-	-	-	2	8,000,000	8,000	Large subsidence chambers
Rockville, Maryland	[a]Traveling grate, 3, 350	-	-	X	X	-	-	-	-	-	3	4,874,973[b]	4,640	Land cost ($297,580) included, completed 1965
Oyster Bay, N.Y.	[a]Continuous feed, 2, 250	-	-	X	X	-	-	-	-	-	2	2,493,000[b]	5,000	Equipped with CCTV
London (Deephams)	1,333 tons/day	Unknown	-	-	-	-	-	-	-	-	-	25,200,000[b]	18,900	Under construction
Paris (Ivry)	[a]Martin, 2, 330,000 tons/ per year	Unknown	X	-	X	X	X	X	Unknown	Unknown	1	30,600,000[b]	Unknown	Under construction

[a]All plants burn mixed, unsegregated refuse.
[b]Not corrected to 1965 index.

Source: PB 200,514 June 1971

to a 1965 index and are shown in Table 1. Costs (uncorrected) of new incinerators either built or being built in Rockville, Maryland; Oyster Bay, New York; London, England; and Paris, France, are also included.

Figure 3.1 and Table 2 illustrate installed capital costs for multiple hearth furnaces.

FIGURE 3.1: MHF INSTALLED CAPITAL COST

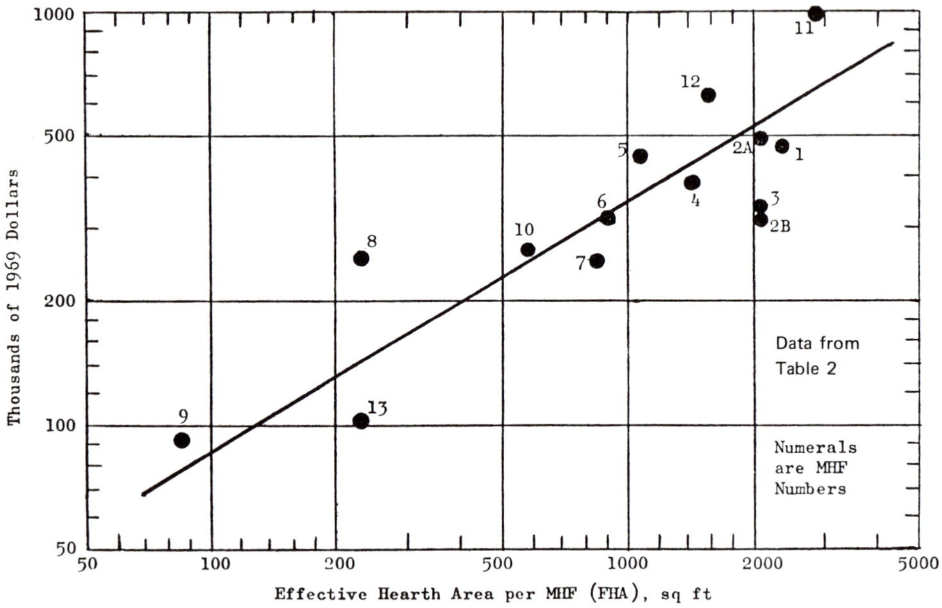

Source: PB 211,264 July 1971

Many authors have given ranges for construction costs per ton dry plant capacity. Table 3 summarizes the ranges of combustion costs per ton dry plant capacity that have been reported in the literature. Foreign costs were directly converted to dollars and do not reflect the differences in purchasing power of the dollars in foreign countries.

The high cost per ton-day of some foreign incinerators is because of their sophisticated gas-cleaning equipment and extensive waste-heat-recovery apparatus. These dual-purpose incinerators not only burn refuse but supply steam and electricity, the value of which more than recovers the extra construction costs.

TABLE 2: MHF INSTALLED CAPITAL COST

MHF No.	Location	(B=Bid) Contract Year	No. of MHF's	Unit MHF Hearth Area, Ft^2	Source for Costs; Remarks (CE=Consulting Engrs.)	Orig. Cost per Unit MHF (1000$)	1969 Cost per Unit MHF (1000$)
1	Cleveland	1963	4	2327	Havens & Emerson (CE)	376	478
2A	Minn.-St. Paul	1938	3	2084	Toltz, King, Duvall, Anderson (CE)	127	496
2B	Minn.-St. Paul	1950	1	2084	1951 M-St.P. Sanitary District Report (19th Annual), p.86	163	314
3	Kansas City	1964	3	2084	Black & Veatch (CE)	247	338
4	Cleveland	1938	4	1425	Havens & Emerson (CE); bulky combustion air preheater included.	99	386
5	Battle Creek	1966	1	1068	McNamee, Porter & Seeley (CE); Elaborate controls included.	380	446
6	Battle Creek	1960	1	890	Malcolm Pirnie & Co. (CE); Estd cost of building subtracted.	241	317
7	Saginaw	1963	1	845	Hubbell, Roth & Clark (CE)	197	250
8	Hatfield	1967	1	230	Tracy Engineers, Inc. (CE)	225	253
9	Bridgeport	1964	1	85	George B. Mebus, Inc. (CE); costs of MHF components available.	74	92
10	South Tahoe	1967	1	575	South Tahoe Public Util. Dist; Estd. dewatering etc. costs subtracted.	233	262
11	Minn.-St. Paul	1965	3	2808	Toltz, King, Duvall, Anderson (CE); Estd dewater. costs subtracted.	816	993
12	Saginaw	1969B	1	1560	Metcalf and Eddy (CE)	630	630
13	East Rochester	1963	1	230	Lozier Engineers (CE)	80	102

Source: PB 211,264 July 1971

TABLE 3: RANGES OF INCINERATOR CONSTRUCTION—COSTS PER TON-DAY

Location	Cost, $
United Kingdom	1,500 to 18,900
W. Europe	6,000 to 16,000
United States	3,000 to 8,000+

Source: PB 200,514 June 1971

AIR POLLUTION CONTROL EQUIPMENT COST

The costs of air pollution control equipment are closely related to the efficiency of the equipment. It is the more sophisticated, high-efficiency equipment that is more costly. The rather simple nonmechanical types of control equipment, such as settling chambers, baffle collectors, and scrubbers do not require pretreatment of the furnace gases and are generally less costly. Scrubbers, however, are perhaps the most expensive of the group.

TABLE 4: APPROXIMATE COSTS OF RECOMMENDED MULTIPLE-CHAMBER INCINERATORS AND SCRUBBERS IN 1968[a]

"Size" of incinerator, lb/hr	General-refuse incinerators	Scrubbers[b]
50	$ 1,200	$ 2,200
100	1,700	3,000
150	2,000	3,600
250	2,700	4,400
500	5,000	6,200
750	9,500	7,600
1000	12,500	8,800
1500	20,000	11,200
2000	25,000	13,200

[a]Incinerator costs are exclusive of foundations.
[b]Scrubber costs are exclusive of foundations but include reasonable utility connections.

Source: PB 190,249 November 1969

Cyclones, electrostatic precipitators, and fabric filters all require pretreatment of the furnace gases and are the most expensive, but most efficient types of control systems. Cyclones are not quite as expensive on the average as fabric filters and electrostatic precipitators. The first electrostatic precipitators to be installed in the United States had an estimated cost of approximately $450,000 each. Table 5 contains estimated construction costs in 1966 of several combinations of air pollution control equipment for a hypothetical 500-ton-per-day incinerator equipped with two 250-ton-per-day furnaces.

Dust loading to the collector was assumed to be 3.5 pounds of dust per 1,000 pounds of flue gas corrected to 50% excess air. Five combinations are given for refractory-lined furnaces and three combinations are given for steam-generating, water-walled furnaces. It can be seen that reduced gas volumes resulting from water-walled furnaces lower the costs of a given control system to less than one-half the value for a refractory furnace.

TABLE 5: COSTS OF CONSTRUCTING, OWNING, AND OPERATING AIR POLLUTION CONTROL EQUIPMENT TO MEET MUNICIPAL INCINERATOR STACK EMISSIONS

Equipment	Average Construction Cost, Dollars	Unit Cost, Dollars/Ton of Refuse Burned	Stack Emissions, Lb/1,000 Lb Flue Gas at 50% Excess Gas
Baffled spray chamber	188,200	0.77	1.75
Spray chamber-cyclone collector	270,360	1.23	0.77
Wet scrubber	400,900	2.10	0.14
Spray chamber-electrostatic precipitator	501,770	1.21	0.175
Spray chamber-fabric filter	712,190	2.00	0.035
Water-cooled furnace-cyclone	91,800	0.38	0.77
Water-cooled furnace-electrostatic precipitator	210,300	0.39	0.175
Water-cooled furnace-fabric filter	243,000	0.65	0.035

Source: PB 200,514 June 1971

REFUSE INCINERATION COSTS

The overall cost of incineration usually consists of personnel, maintenance, repair, replacement, utility, and amortization costs. Computations of cost in man-hours per ton of refuse for four European and six incinerators in the United States are given in Table 6. All of the European incinerators are equipped with by-product recovery and salvaging operations for which extra man-hours are required.

A comparison of the overall costs from this limited data strongly indicates that, on the average, European operating costs are substantially lower than operating costs in the United States. This is even more indicative when income from by-product recovery and salvage operation is included.

TABLE 6: PERSONNEL REQUIREMENTS OF PLANTS BURNING MIXED, UNSEGREGATED REFUSE

Plant	Design capacity per plant, tons/day	Operating period, hr/week	Operating factor[a]	Refuse processed, tons/working day	Manpower, man-hours/working day	Cost/man-hr-ton, $	Cost,[b] dollars/ton
			European				
A	600	168	70	420	384	0.91	–
B	1,000	168	85	850	564	0.66	–
C	1,540	126	70	980	592	0.60	2.50 net
D	400		50	200			3.50 gross 1.50 net
			U.S.A.				
6 modern continuous-feed plants[c]	1,000	128	80	800	576	0.72	4.30

Note: All European incinerators are equipped and operated to include steam and power generation, residue processing, and metal salvage.

[a]Operating Factors = Average actual production ÷ Design capacity.
[b]Total cost exclusive of amortization.
[c]Average values per plant for 6 modern plants for 3 years of 300 working days each—refuse incineration only.

Source: PB 200,514 June 1971

EXPANSION AND REMODELING COSTS

Substantial savings can be realized by remodeling and expanding some of the older incinerators in lieu of replacement with new installations. Many of the recently constructed incinerators have allotted space for the installation of additional furnaces and air pollution control equipment. In such an installation, there is no question as to whether it is more economical to expand and remodel or to build a new plant. It appears from the published literature that rehabilitating old plants can be economically attractive.

The New York Department of Sanitation operated 11 incinerators that were built between 1934 and 1962. Four of these were batch-feed units built before 1938. Two of the batch-feed furnaces were to be converted to continuous-feed furnaces and the remaining 7 continuous-feed incinerators were being rehabilitated. Construction costs for sophisticated air pollution control systems were estimated to range from $1 to $1½ million.

The city of Philadelphia has considered the conversion of some of its batch-feed incinerators to a continuous-feed type. The city operated two continuous-feed and four batch-feed incinerators. Rehabilitation and conversion of the batch-feed incinerators were estimated to cost $2.1 to $2.2 million. Replacement of one of the batch-feed units may cost approximately $6 to $7 million, assuming a capacity of 600 tons per day.

The city of Kenosha, Wisconsin, has studied the possibility of converting its city incinerator plant from a garbage-only incinerator operation to a modern, mixed-refuse incineration plant. An entirely new 240-ton-per-day incinerator plant, including air pollution control equipment, would cost approximately $1.7 million, but cost estimates for rehabilitation of the old plant range from only $333,200 to $661,900, depending on design features and air pollution control equipment.

BY-PRODUCT RECOVERY

Waste heat, metals, slag, and residue can all be valuable by-products of municipal incineration. Refuse as a fuel is quite attractive. It is free and, in contrast to other fuels, contains small amounts of sulfur. Nearly 2 pounds of steam can be generated per pound of refuse incinerated. The new Ivry plant built in Paris, France, will produce 277,000 pounds of steam per hour at 1,378 pounds per square inch and 878°F. Thus a rather significant income can result from the sale of steam or electricity to a nearby market. A modern refuse incinerator in Germany receives approximately $1.99 per ton of refuse incinerated from the sale of steam. Sale of steam from a 720-ton-per-day incinerator in Chicago, Illinois, was estimated to net $125,000 to $150,000 annually.

Waste heat recovery on a smaller scale for internal use only by the incinerator plant is fairly common throughout the United States. This type of operation can be reflected in significant savings in utility bills. The use of incineration waste heat for the desalinization of salt water is another cost recovery method for incinerators located in coastal areas. The Oceanside Refuse Disposal Plant in New York desalts ocean water for internal use. Table 7 summarizes the uses of waste heat for 43 municipal incinerators in the United States. Each use, except the preheating of combustion air, has direct economic value resulting in lower total incineration cost.

TABLE 7: INCINERATOR WASTE HEAT UTILIZATION

Construction year	Plants reporting use	Heat use								
		Building heat and/or hot water	Electric power	Sewage sludge drying	Steam production				Preheat combustion air	Other use
					For sale	Outside heating	Other use	Use not stated		
1945–1950	2	1	1	0	0	0	0	0	0	0
1951–1955	10[a]	4	2	2	1	0	0	2	1	0
1956–1960	17[b]	10	1	3	0	1	d,g	1	1	e,f
1961–1965	14[c]	9	1	1	1	1	h	1	0	i
Total	43	24	5	6	2	2	3	4	2	3

[a] One plant reports building heat, hot water, and preheating combustion air. Another reports building heat and sludge drying.

[b] One plant each reports: Hot water and power generation, hot water and air preheating, hot water and sludge drying, and steam for equipment drives and heating nearby hospital.

[c] One plant reports building heat and steam for sale. One reports power generation and desalination.

[d] Equipment drives.

[e] Sludge furnace.

[f] Heat for sludge digester.

[g] For sewage treatment plant.

[h] Desalination of seawater.

[i] Tubular gas reheater cools combustion-chamber outlet gas and reheats scrubber exit gas.

Source: PB 200,514 June 1971

Salvage of residue and metal is perhaps more widely practiced in foreign countries than it is in the United States. The price received for salvage metal in Europe and England, prior to 1966, ranged from $2.15 to $18.50 per ton. Similar variations in the price of salvage metals in the United States are common.

With increasing fly ash collection in municipal incineration, sintering operations to produce building blocks, such as those found at some power companies, may find wider application. Further profitable uses of residue and fly ash are the objectives of some current research programs. Some incinerator salvage operations in the United States have had to be abandoned because they were not economically practical. Many incinerators have successful salvaging operations, however, such as one incinerator in Chicago, Illinois, that burns 500 tons of refuse per day, of which approximately 170 cubic yards of reclaimable metal is sold daily.

Based on one year of operational data, 38 cents worth of metal per ton of raw refuse has been recovered. Tin cans accounted for 23 cents per ton and the remaining 15 cents per ton is received from the sale of scrap metal.

Cinders are sold for fill, and the remaining residue can be mixed with lime to form a product that is used in the construction of streets, parking lots, and playgrounds. This product, even though proven useful, has not been financially successful.

USE OF AN ACCOUNTING SYSTEM FOR INCINERATOR OPERATION

The material in the following section is taken from a 1970 report written by E.R. Zausner for the U.S. Department of Health, Education and Welfare.

Effective solid waste management requires an adequate information system including data on activity and the costs of operation and ownership. Although a cost accounting system represents only one part of the total system, it does facilitate the collection and later utilization of the data obtained.

Present information on incineration and its associated costs is both inadequate and nonstandardized. A cost accounting system can aid in controlling the costs and performance of its incinerator operations, as well as aid in formulating future plans.

Cost Centers and Cost Allocations

The complexity of incinerator operations requires a breakdown and description of operations to facilitate analysis. The incinerator is assumed here to consist of several interrelated suboperations, each of which is analyzed separately. These suboperations are called cost centers because costs are accumulated separately for each of the major functional activities. Analysis and control are simplified if excessive costs or inefficiencies can be traced to a functional activity or area of the facility.

The number of cost centers increases as the size and complexity of operations increase. More cost centers, however, require the collection of more data and, therefore, increase costs. For most facilities, four cost centers appear to result in adequate data without incurring excessive data acquisition costs.

Three of the four cost centers (Receiving and Storage, Volume Reduction, and

Effluent Handling and Treatment) are termed the direct cost centers because they can be directly associated with certain incinerator operations and unit processes. The operations included in each follow the process flow from input of raw wastes to output of effluents (Figure 3.2). The fourth, the Repairs and Maintenance cost center, cannot be directly associated with waste processing. Therefore, it is separated from other operations and not shown in the diagram. Because it incurs a large percentage of operating costs, a separate analysis is needed.

Larger operations may require more cost centers. For instance, the Effluent Handling and Treatment cost center could be divided into Air Pollution, Water Treatment, and Residue Handling cost centers. Similarly, salvage or heat utilization operations should be put in separate cost centers.

These cost centers classify the operations by function. The costs incurred are for labor, parts and supplies, utilities, and overhead, and they must be allocated to the cost centers in an accurate and representative manner (Figure 3.3). Note that costs are first allocated to all four cost centers; the Repairs and Maintenance cost center is then allocated to the three direct cost centers. The result is the total operating cost for each direct cost center.

There are many alternatives for actually allocating the operating costs. A straight-forward method for each type of expense will be outlined. Labor costs may be allocated to the four cost centers based on the relative number of hours employees worked in each area and their respective wage rates. Utilities may be allocated based on an engineering estimate of the relative usage rates of the equipment in each cost center. Both water and electricity should be allocated.

Parts and supplies will be allocated to each direct cost center after first being recorded in the Repairs and Maintenance cost center. General overhead, which includes supervision, insurance, etc., can be allocated equally to each cost center or on the basis of the relative number of employees in each cost center. The latter technique is recommended. Finally, the Repairs and Maintenance cost center is allocated to the three direct cost centers based on the actual expenses incurred in each one.

The sum of the costs of these three direct cost centers is the total operating cost. The total annual cost of operations represents these operating costs plus the costs of financing and ownership. The actual forms are designed to facilitate the collection and later allocation of costs to these cost centers.

Forms and Reports

The reports are most easily grouped into those that are used to collect the data on operations and those used to reduce and present the data for the purposes of analysis, decision making, and control.

This data reduction and presentation cannot be accomplished without the daily

FIGURE 3.2: INCINERATOR COST CENTERS

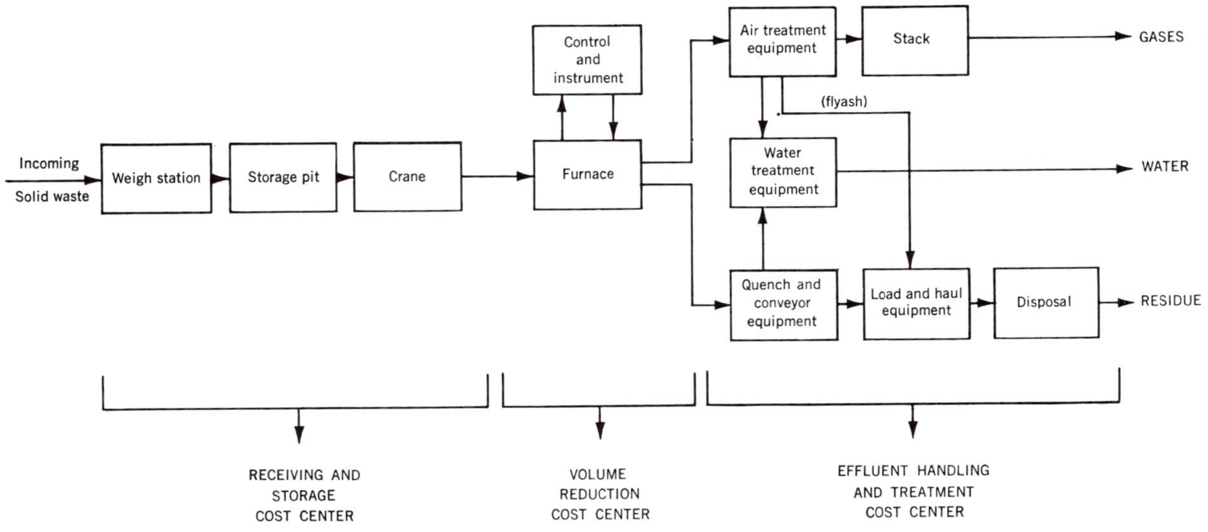

Source: Public Health Service Publication 2032, 1970

FIGURE 3.3: ALLOCATION OF COSTS

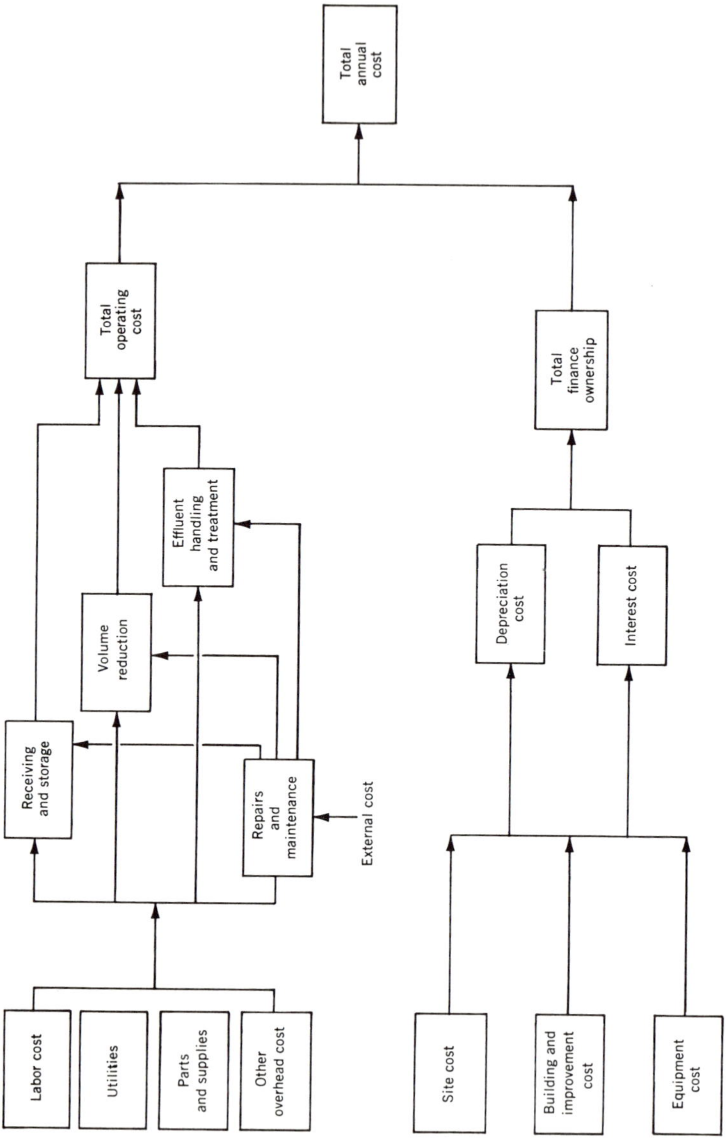

Source: Public Health Service Publication 2032, 1970

recording of all pertinent activity and cost information. Data not recorded daily are not retrievable at some later date. Incinerator personnel, supervisors, and others involved in operations primarily use the following forms (1 through 4) to record the data required.

Weekly Labor Report (Form 1, Figure 3.4): Daily entries of labor activity are recorded in duplicate at the site. One copy is forwarded to the payroll department for determining weekly wages. The incinerator supervisor and the accounting department use the other copy for computing total labor hours and assigning these hours and associated costs to the four cost centers.

Daily Truck Record (Form 2, Figure 3.5): The waste received and residue removed, as well as the types and sources of waste received, are recorded manually on this form for the entire day. (If the incinerator has a scale that automatically records the weight information, that part of the form would be replaced by the weight ticket or record of the scale.) Each delivery is recorded separately by the weighmaster. A second weighing of the empty truck may be taken or the vehicle's tare weight (as officially determined by a licensing agency, etc.) may be substituted.

The form is forwarded to the accounting department at the end of each month. In addition to utilizing recorded weight data to bill private users later, the sources and types of waste data are useful in special analyses of trends, compositions, and distributions of solid wastes in the community.

Daily Report on Incinerator Operations (Form 3, Figure 3.6): When there is actual downtime and repairs are required, the expenses that will be allocated to the three direct cost centers are recorded on the lower half of the two-purpose report. These data are particularly useful in analyzing equipment performance and cost. In addition, data on utility usage are recorded on the form at the end of each month.

The top of the report is used to summarize the daily operations. The employee and activity data give management personnel who are not at the site daily, but who still require daily feedback on operations, a quick and accurate summary of the day's activities. The performance data are also useful in assessing daily efficiency. The report is completed daily, sent to the main office, and filed for later use.

Incinerator Capital Investment Report (Form 4, Figure 3.7): This form is completed when construction is finished or when the cost system is first implemented. Only when improvements or new equipment are either constructed or purchased is it updated. In addition to collecting the data required to calculate depreciation for the period and allocating it to cost centers, the form also summarizes the bond and interest information required to compute the total costs of financing and ownership.

For the most part, Forms 1 through 4 are utilized to collect the data associated

FIGURE 3.4: WEEKLY LABOR REPORT (FORM 1)

INCINERATOR: ______________________ DATE: ____/____/____

SIGNATURE: ______________________ SHIFT: ______________________

Employee ident.	Day 1		Day 2		Day 3		Day 4		Day 5		Day 6		Day 7		Individual totals	Comments (Note causes and hours of absence, etc.)
	Cost center	Hours	Cost center	Hours	Cost center	Hours	Cost center	Hours	Cost center	Hours	Cost center	Hours	Cost center	Hours		
Totals	X		X		X		X		X		X		X			XXXXXXXXXXXXXXX

Instructions: Incinerator supervisor to complete this form daily. List all employees separately including temporary help. "Hours" refers to hours worked daily. At the end of each week forward one copy to the payroll department and retain the original for further use.

Abbreviations of cost centers and workers to be assigned to each: R&S = receiving and storage: crane operator, weightmaster, tipping floor, and charging attendants. VR = volume reduction: stokers, control monitors, etc. EHT = effluent handling: residue haulers, disposal site operators, etc. R& M = repairs and maintenance: include all general maintenance workers and part time repairmen.

Source: Public Health Service Publication 2032, 1970

FIGURE 3.5: DAILY TRUCK RECORD (FORM 2)

INCINERATOR: ____________________ DATE: ____________________

SIGNATURE: ____________________ SHIFT: ____________________

No.	Truck * identification	Time	Wastes		Weight in	Weight out (or tare weight)	Net amount	
			Source	Type ‡			Wastes	Residue
1								
2								
3								
4								
5								
6								
7								
8								
9								
10								
11								
12								
13								
12								
15								
16								
17								
18								
19								
20								
Totals		X	X	X	X	X		

Instruction: To be completed by weighmaster for each delivery of waste or removal of residue.
* Truck identification is number of the public truck; if private vehicle, the name of company for billing purposes.
† Source: R = residential; C = commerical; I = industrial.
‡ Type: R = rubbish; G = garbage (also unusual items).

Source: Public Health Service Publication 2032, 1970

FIGURE 3.6: DAILY REPORT ON INCINERATOR OPERATIONS (FORM 3)

SITE: ________________________ DATE: ______________

PERFORMANCE DATA
% Weight reduction $\left(\frac{\text{residue}}{\text{wastes burned}}\right)$:
Man hours per ton:
Tons of residue per trip:
Number of injuries:

EMPLOYEE HOURS				
Cost center	Shift			
	1	2	3	4
Receiving				
Volume reduction				
Effluent				
Repairs				
Totals				

ACTIVITY DATA		
	Loads	Tons
Wastes received		
Wastes burned		
Left in pit		
Residue		

REPAIRS AND MAINTENANCE DATA								
Equipment description	Cost center	Cause	Hours down	Labor hours	Labor cost	Parts cost	External costs	Total cost

UTILITY DATA (Only complete this section at the end of the month)			
	Electric	Gas	Water
Meter reading			

Source: Public Health Service Publication 2032, 1970

FIGURE 3.7: INCINERATOR CAPITAL INVESTMENT REPORT (FORM 4)

INCINERATOR: ______________________ DATE: ____________

Description	Size, capacity, amount, etc.	Date put in use	Estimated total life	New cost	Other comments	Yearly depreciation	Monthly depreciation
Site:							
Land						X	X
Surveys							
Preparation							
Roads							
Other							
Buildings:							
Scale house							
Pit							
Offices							
Main building							
Stacks							
Other							
Equipment:							
Scales							
Crane(s)							
Furnace(s)							
Air pollution							
Water treatment							
Residue removal (including vehicles)							
Instrumentation and control							
Other							
Totals	X	X	X		X		

FINANCING DATA

Bond type	Face value	Premium or discount	Interest rate	Yearly interest *	Monthly interest

Instructions: To be completed by supervisor or accounting department. Depreciation may be straight-line or on an accelerated basis.

* Interest must account for net effect of premium or discount on bond sale.

Source: Public Health Service Publication 2032, 1970

with the construction and operation of an incinerator. The cost of accumulating these data can only be justified by its intensive and effective utilization. This is accomplished by meaningful data reduction and presentation. The data must be presented clearly and quickly to the personnel who can use it most effectively for analyses and control. The following forms (as well as Form 3) are designed to fulfill these objectives.

Incinerator Operations Summary (Form 5, Figure 3.8): This form summarizes six distinct groups of information about incinerator operations for a specific period. For control purposes, monthly reports would be desirable, although less frequent preparation would be possible. The first two segments present activity and operating cost data for the total operation. Costs are broken down by type to aid the cost analysis, and the activity and performance factors are designed to help analyze inefficiencies and performance. The remaining four sections break the costs into the four cost centers.

Total operating costs are presented for each area as are other factors that may be useful to analyze the functional activities. Obviously, there are many other factors and costs that could be presented. The ones illustrated, however, are adequate for most analyses. Nonetheless, modifications or additions should be made for facilities with different operations and data requirements.

This form, designed for control purposes, contains only controllable expenses for which the supervisor can be held accountable; capital or financing costs are not included. The form is prepared by the accounting department from the data in Forms 1, 2, and 3 and additional data on file concerning labor rates, insurance, fringe benefits and charges from other departments, external expense billings, etc. Copies of the form are forwarded to both the facility supervisor and to his superior. Analysis of the form indicates excessive expenses and aids the supervisor in taking corrective action.

Incinerator Total Cost Report (Form 6, Figure 3.9): All the activities and costs incurred by the incinerator during the period are summarized from data in present and past Incinerator Operations Summaries (Form 5) and from the depreciation and interest data available in the Incinerator Capital Investment Report (Form 4). Semiannual and annual preparation would be sufficient. Form 6—Alternate (Figure 3.10) can be used if disposal charges or other types of revenues are associated with incinerator operations.

Summary of Information Flow

Operating data are accumulated daily at the incinerator site and transmitted periodically to the accounting department. The accounting department combines these reports with additional information it accumulates to get total operating costs. This summary is then returned to the supervisor for his own use. Next, the accounting department combines the operating cost data with the depreciation and interest cost data (from the Incinerator Capital Investment Reports) to compute total costs for the period. This total cost information is

FIGURE 3.8: INCINERATOR OPERATIONS SUMMARY (FORM 5)

INCINERATOR: REPORT PERIOD: from to

ACTIVITY AND PERFORMANCE

	Actual amount	± % Budget variance
Tons incinerated		
% Weight reduction		
Total labor hours		
% Capacity utilized		
% Capacity available		
% Utilized/% available		

OPERATING COST TOTALS

	Actual amount	± % Budget variance
Total operating cost		
Total labor cost		
Utilities cost		
Parts and supplies		
Outside charges		
Overhead		

RECEIVING COSTS

	Actual amount	± % Budget variance
Total operating cost per ton		
Labor hours		

VOLUME REDUCTION COSTS

	Actual amount	± % Budget variance
Total operating cost per ton		
Labor hours		
Average operating temperature		

EFFLUENT HANDLING COSTS

	Actual amount	± % Budget variance
Total operating cost per ton		
Gallons of water per ton		
Tons of residue per load		

REPAIRS AND MAINTENANCE COSTS

	Actual amounts	± % Budget variance
Total operating cost		
Receiving repair costs		
Volume reduction repair costs		
Effluent handling repair costs		

Source: Public Health Service Publication 2032, 1970

FIGURE 3.9: INCINERATOR TOTAL COST REPORT (FORM 6)

SITE: ______________________ REPORT PERIOD from ________ to ________

Data	For this period	±% Variance from budget for this period	Year to date	±% Variance from budget for year to date
Tons incinerated				
Weight reduction				
Total operating cost				
Total financing and ownership cost				
Total cost				
Operating cost per ton				
Financing and ownership, cost per ton				
Total cost per ton				

Instructions: To be completed by accounting department from data available in "Incinerator Operations Summary" and "Incinerator Capital Investment Report" when requested or periodically. Copy sent to city manager, head of department of public works, or their equivalent.

Source: Public Health Service Publication 2032, 1970

FIGURE 3.10: INCINERATOR TOTAL COST SUMMARY (FORM 6–ALTERNATE)

SITE: ____________ REPORT PERIOD: from ________ to ________

Data	For this period	Budget—this period	Year to date	Budget—year to date
Tons of waste incinerated				
Percent weight reduction				
Total operating cost				
Total financing and ownership cost				
Total cost				
Operating cost per ton				
Financing and ownership cost per ton				
Total cost per ton				
Revenues—other communities				
Revenues—private collectors				
Revenues—miscellaneous				
Total revenues				
Total revenues per ton				
Net cost (profit)				
Net cost (profit) per ton				

Source: Public Health Service Publication 2032, 1970

then given to the heads of departments of sanitation and public works, or their equivalents.

System Utilization

Only with efficient and intensive utilization of the information generated from the accounting system and forms can the additional time, effort, and money required to implement and maintain the system be justified. The system's intensive use promotes two major objectives: quality control and cost control. Reduced costs must be accomplished without deteriorating and operating quality. Similarly, quality is interrelated with the costs of obtaining it.

All the factors that affect the quality and effectiveness of incinerator operations can be translated into costs. Amount of volume reduction, residue characteristics, and the levels of stack emissions and water pollution determine the quality of operations. Cost control does not call for economizing at the expense of quality. On the contrary, once a level of acceptable operation has been determined along with the attendant costs, the cost control system can help the supervisor maintain that level of operation.

Effective cost control requires timely recognition of excessive costs and identification of responsibility for the increased costs. Comparing units costs (cost per ton of waste incinerated) with both the current budget and the corresponding period last year helps indicate excessive costs. The use of unit cost facilitates the analysis of costs, independent of changes in the level of activity. The cost-center breakdowns help single out the responsible factor or person. This system allows both of these critical factors to be determined; corrective action may then be effectively initiated.

The Incinerator Total Cost Report (Form 6) can indicate to the highest level of municipal management, i.e., the city manager or the head of the sanitation department, if costs are excessive. If so, the supervisor of the particular facility can be held responsible to the extent that his operating costs have increased. The supervisor, in turn, can analyze the cause of this cost rise. He may trace the increased cost to the type of cost, as well as the cost center, and possibly to the employee or piece of equipment responsible. All of the needed data are in Form 5, the Incinerator Operations Summary.

Incinerator Emissions

Emissions from incinerators will greatly contribute to the air pollution problem, unless they are effectively reduced by improved design or air pollution control devices now available on the market. In a study prepared for the EPA and published as PB 210,382, February 1972 by Arthur D. Little, Inc., data is presented (Table 1) which indicates the magnitude of the incinerator air pollution problem both now and in the future.

These emission projections show clearly the important contribution to total emissions of combustible gaseous and particulate pollutants. Previous work has shown that the very existence of these pollutants in incinerator flue gas is a consequence of inadequate mixing within the furnace and flues of the incinerator.

TABLE 1: ESTIMATED INCINERATOR EMISSIONS*

	1968 Emissions Estimate*			2000 Emissions Estimate*		
	Furnace	Stack	%	Furnace	Stack	%
Combustible Particulate	38	32	7.1	131	49	3.3
Carbon Monoxide	280	280	62.2	829	829	56.0
Hydrocarbons	22	22	4.9	64	64	4.3
Polynuclear Hydrocarbons	0.01	0.005	0	0.03	0.0009	0
Subtotal (Combustibles)	340	334	74.2	1,024	942	63.6
Mineral Particulate	90	56	12.4	708	118	8.0
Sulfur Dioxide	32	32	7.1	161	160	10.8
Nitrogen Oxides	26	22	4.9	147	114	7.7
Hydrogen Chloride	8	6	1.3	219	147	9.9
Volatile Metals	0.3	0.3	1.0	0.055	0.025	0
Total	496	450	100.0	2,259	1,481	100.0

*Thousands of tons per year.

Source: PB 210,382 February 1972

Specifically, theoretical and experimental studies indicate that at the temperature and air-fuel conditions which are obtainable in well-mixed incinerators, the majority of combustible pollutants are destroyed within a small fraction of the typical gas and particle residence times. Thus, if means can be found to assure that complete mixing occurs within incinerator furnaces, the emission levels of most combustible pollutants can almost be eliminated.

The technology supporting municipal incinerator design, however, has evolved slowly and is yet immature. Indeed, there are few design parameters which have been well-characterized and proven useful in practice. Most often, the only measures of design adequacy have been such pragmatic "variables" as plant availability or the lack of maintenance and operating headaches. In the areas of combustion efficiency and air pollution performance, there is almost no guidance for the designer, only the documented acts of unacceptably high emissions.

PARTICULATE EMISSIONS

Particle Size

Particulate matter constituents that have been identified in municipal incinerator effluents include smoke, soot, flyash, grit, carbonaceous flakes, aldehydes, organic acids, esters, fatty materials, phenols, hydrocarbons, and polynuclear hydrocarbons. The ease with which the large particles break up during and after capture makes their measurement difficult. Limited data has been published on incinerator emissions.

Table 2 gives the breakdown of particle size and other physical properties of particulate matter gathered in the area between the combustion chambers and the gas-cleaning devices for three incinerators, each with a different grate system. Furnace capacity ranged from 120 to 250 tons per day. Tests performed on another incinerator having a furnace capacity of 150 tons per day provide particle size data for stack gases that had been previously cleaned by a combustion settling chamber and wet baffle system. The sizes of particles measured by a Coulter Counter are presented in Table 3.

Particle sizes recommended for proper application of electrostatic precipitators to European incinerators are given in Table 4. In comparing Tables 3 and 4, keep in mind that Table 3 reports actual measurements from an incinerator stack in Milwaukee, Wisconsin, where the stack gases had been previously cleaned by a settling chamber and a wet baffle system. Table 4 presents typical values, based on European data, of particle sizes that can be used for design of electrostatic precipitators. The particulate sizes in Table 4 are representative of those particles that have passed through a waste-heat-boiler system at some point just prior to entry into the precipitator.

TABLE 2: PHYSICAL PROPERTIES OF PARTICLES LEAVING FURNACE

	---- Installation Number ----		
	1	2	3
Total sample in cyclone, %	77.0	77.5	63.0
Total sample in bag, %	23.0	22.5	37.0
Specific gravity, g/cm^3	2.65	2.70	3.77
Bulk density, lb/cf	-	30.87	9.4
Loss on ignition at 750°C, %	18.5	8.15	30.4
Analysis			
% by weight less than 2 microns	13.5	14.6	23.5
% by weight less than 4 microns	16.0	19.2	30.0
% by weight less than 6 microns	19.0	22.3	33.7
% by weight less than 8 microns	21.0	24.8	36.3
% by weight less than 10 microns	23.0	26.8	38.1
% by weight less than 15 microns	25.0	31.1	42.1
% by weight less than 20 microns	27.5	34.6	45.0
% by weight less than 30 microns	30.0	40.4	50.0

TABLE 3: SIZE AND DENSITY OF INCINERATOR STACK GAS PARTICLES*

Microns	% By Weight Greater than Stated Size
30	31.3
20	52.8
10	79.5
5	94.0

*Density is 1.85 grams per cubic centimeter.

TABLE 4: PARTICLE SIZE AND DENSITY FOR DESIGN OF ELECTROSTATIC PRECIPITATORS IN EUROPEAN INCINERATORS

Particle Diameter, microns	Percent
$<$ 5	11.1
$<$ 7	21.7
$<$10	30.1
$<$14	40.2
$<$19	46.9
$<$27	55.1
$<$39	67.5
$<$59	90.5
$>$59	9.5

Source: PB 200,514 June 1971

Expressing Particulate Concentration

Particle concentrations are normally expressed in one of three ways. Dust loading, a term frequently used, expresses particulate concentration in grains (7,000 grains equals 1 pound) per cubic foot of gas normally corrected to either 12% CO_2 or 50% excess air. A second method expresses dust loading in pounds per 1,000 pounds of gas corrected to either 12% CO_2 or 50% excess air. A third method is to express concentration in terms of weight of particulate matter emitted per weight of refuse burned. Carbon dioxide produced by the use of auxiliary fuels should be excluded from the calculation to 12% CO_2. Samplings of particulate matter can be taken at various gas temperatures and atmospheric pressures. To make various samples comparable, a correction to standard conditions of 68°F and 29.92 inches of Hg can be accomplished.

Particulate Concentration

In an EPA Publication, PB 216,372 (1970), prepared by W.C. Achinger and L.E. Daniels of the Bureau of Solid Waste Management, particulate emission data is given which reflects incinerator design, operation, and air pollution control equipment for several incinerators (Table 5). These incinerators are described below.

> *Incinerator A* was built in 1966 with a design capacity of 300 tons per day. Each furnace contains a modified reciprocating grate and a stationary grate. A wetted-column water scrubber is used for air-pollution control.
>
> *Incinerator B* was built in 1966 with a design capacity of 300 tons per day. Each furnace contains three sections of rocking grates. The air-pollution-control system is a flooded, baffle-wall water scrubber.
>
> *Incinerator C*, built in 1967, is a pilot-plant conical burner with a design capacity of 1,000 lb/hr. A centrifugal water scrubber, an afterburner, an electrostatic precipitator, or some combination thereof is used for air-pollution control.
>
> *Incinerator D* was built in 1965 with a design capacity of 500 tons per day. Each furnace contains two sections of traveling grates, and a flooded, baffle-wall water scrubber is used for air-pollution control.
>
> *Incinerators E and F* are rotary kilns built in 1963 with design capacities of 500 and 600 tons per day, respectively. Each furnace contains three sections of reciprocating grates followed by a rotary kiln. Air-pollution control is achieved through a baffle-wall and water-spray system.
>
> *Incinerator G* was built in 1967 with a design capacity of 400 tons per day. Each furnace contains 4 sections of reciprocating grates. A multitube dry cyclone following a wet-baffle wall is used for air-pollution control.

TABLE 5: PARTICULATE-EMISSION DATA

	Particulate Emissions								
Incinerator	gr/st ft^3 at 12% CO_2	lb/1000 lb at 50% Excess Air	lb/hr	lb/ton of Waste Charged	CO_2 %	Stack Temp. (°F)	Excess Air (%)	Moisture (%)	Gas-Flow Rate (ft^3/min)
A	0.55	1.06	122	10.4	4.6	455	270	16.3	69,800
B	1.12	-	186	14.5	3.5	585	-	16.1	131,000
C-1	0.56	0.75	3.2	4.1	2.8	138	370	18.0	3,890
C-2	0.41	0.46	2.4	3.4	3.3	158	220	25.6	3,990
C-3	0.30	0.52	1.7	2.9	3.3	325	410	15.5	4,460
D	0.46	0.85	173	8.8	5.0	485	260	18.1	120,000
E	0.73	1.19	238	8.6	5.0	305	220	26.6	186,000
F	0.72	1.18	-	12.5	3.9	365	320	16.0	165,000
G	1.35	2.70	386	20.4	3.2	500	500	14.3	130,000

Source: PB 216,372 (1970)

In tests conducted on a small, refractory lined, multiple chamber prototype incinerator using synthetic refuse and described in a Public Health Service publication PB 215,093, 1965, A.H. Rose, Jr., et al, concluded that significant effects on particulate could be observed by varying (1) percent excess combustion air, and (2) percent underfire air.

Little or no effect on particulates was observed with an increase of underfire air at the 150% excess air level; but at the 50% excess air level an increase in underfire air from 15 to 30% (of the total combustion air) caused a 210% increase in particulate loadings (both in lb particulate/hr and in lb particulate/lb of fuel charged). Underfire air affected particulate loadings at only one of the two excess combustion air levels selected.

There were no significant effects on particulates attributable to (1) fuel charging rate, (2) stoking interval, (3) percent secondary air, and (4) pounds of fuel per charge.

The fact that combustion of fuel with 150% excess combustion air at 30% underfire produced appreciably lower particulate loadings than combustion of fuel with 50% excess combustion air at 30% underfire, suggests the need for investigating the effects of high underfire air levels involving excess combustion air at levels of 150% and above.

One of the common concepts regarding particulate discharge from municipal and industrial incineration processes is that total solid emissions result solely from physical entrainment of particles in the gas stream with particulate discharge increasing as gas velocity through the fuel bed increases. The concept of physical entrainment of solids as the sole mechanism for producing total solid emissions however is not valid. While physical entrainment probably is a mechanism which contributes to the solid discharges, other factors such as volatilization and condensation of inorganic compounds and incomplete combustion also affect the degree of this discharge.

Particulate Concentration at Different Points

Measurements of particulate emissions are usually made at some point within the incinerator or in the stack. Air pollution control equipment manufacturers are concerned with particle concentrations at the combustion chamber outlets or at the location of the entrance to their gas-cleaning equipment. They are also interested in stack concentrations which indicate collection ability. Air pollution control authorities are primarily interested in the stack emissions rather than internal particulate measurements.

Particle Measurements at Furnace Outlet: Tests have been performed at the furnace outlets of several types and sizes of incinerators. Table 6 summarizes particulate loadings corrected to 50% excess air. The method of supplying air to the combustion chamber is one of the major factors affecting the creation of flyash. This factor is indeed apparent in Table 6 which shows, as previously found in experimental incinerators, that the greater the percentage of underfire air supplied, the higher the particulate loading.

TABLE 6: PARTICULATE EMISSIONS AT FURNACE OUTLET

Furnace Type	Excess Air, %	Underfire Air, %	Average Dust Loading, lb/ton of Charge
50 ton per day batch	235	20	0.78
50 ton per day batch	110	50	1.04
50 ton per day batch	100	70	1.79
250 ton per day continuous	190	20	3.8*
250 ton per day continuous	180	50	2.8
250 ton per day continuous	150	100	4.6
250 ton per day continuous (traveling grate)	(6.0% CO_2)	41.8 scfm/sq ft grate area	12.4
250 ton per day continuous (reciprocating grate)	(5.0% CO_2)	105 scfm/sq ft grate area	25.1
120 ton per day continuous (rocking grate)	(7.0% CO_2)	17.5 scfm/sq ft grate area	9.1
150 ton per day continuous (rocking grate)	-	-	30.8

*See *Municipal Refuse Disposal*, 2nd ed., Chicago, American Public Works, 1966, p. 528, for a special discussion and explanation of this value.

Source: PB 200,514 June 1971

European design and performance factors give a dust loading of from 0.8 to 4.0 grains per standard cubic foot at the entry point of the combustion gases into an electrostatic precipitator. At this point, the gases have been cooled to 400° to 650°F by the waste-heat-recovery boilers. Some of the particles settle out or collect on the boiler tubes; this range of values is, therefore, not truly representative of the dust loading at the furnace outlet.

Stack Emission Measurements: Some stack particulate matter measurements for incinerators equipped with various types of cleaning systems are presented in Table 7.

TABLE 7: PARTICULATE MEASUREMENTS OF STACK GASES

Furnace Type	Control Equipment	Excess Air, %	Underfire Air, %	Average Dust Loading*
50 ton per day batch	Scrubber	235	20	0.57
50 ton per day batch	Scrubber	110	50	0.55
50 ton per day batch	Scrubber	100	70	0.61
250 ton per day continuous (reciprocating grate)	Settling chamber and wet baffle	-	-	11.8
120 ton per day continuous (rocking grate)	Settling chamber and wet baffle	-	-	8.2
150 ton per day continuous (rocking grate)	Wet baffle	-	-	8.24

*lb/ton charge in stack gases.

Source: PB 200,514 June 1971

Stack emission measurements are obviously very closely related to the excess air supplied to the furnace and the efficiency of the gas-cleaning equipment. Measurements of stack emissions are necessary to determine compliance with local codes and to check efficiency of the air pollution control equipment. The samples are generally taken from some convenient stack location.

Particulate Chemical Composition

Little data has been published on chemical composition of incinerator particulate matter. Flyash can consist of an average of from 5 to 30% organic matter and from 70 to 95% inorganic matter. A chemical analysis that gives the various inorganic constituents of incinerator flyash from the South Shore incinerator in New York City is presented in Table 8.

TABLE 8: CHEMICAL ANALYSIS OF FLYASH SAMPLES FROM SOUTH SHORE INCINERATOR, NEW YORK CITY, BY SOURCE*

	Source of Sample		
Component	Upper Flue	Expansion Chamber	Emitted
Organic	0.5	0.6	10.4
Inorganic	99.5	99.4	89.6
Silica as SiO_2	50.1	54.6	36.1
Iron as Fe_2O_3	5.3	6.0	4.2
Alumina as Al_2O_3	22.5	20.4	22.4
Calcium as CaO	7.9	7.8	8.6
Magnesium as MgO	1.8	1.9	2.1
Sulfur as SO_3	4.3	2.3	7.6
Sodium and potassium oxides	8.1	7.0	19.0

*Percent by weight.

Source: PB 200,514 June 1971

A rather detailed elemental analysis of ashed incinerator stack effluent and collector catch is summarized in Table 9. The incinerator tested was equipped with an impingement wet baffle system. Results of two test runs are presented. Stack effluents were found to have a pH of 7.7 and 8.3; a much higher pH of 12.3 was found for the collector catch.

TABLE 9: SPECTROGRAPHIC ANALYSIS OF ASHED INCINERATOR PARTICULATE MATTER

Element	Stack Effluent, % Ashed Material	Collector Catch, % Ashed Material
Silicon	5+ -	10+ -
Manganese	0.1 - 1.0	0.1 - 1.0
Chromium	0.1 - 1.0	0.1 - 1.0
Nickel	1.0 - 10+	0.001 - 0.01
Copper	0.1 - 1.0	0.01 - 0.1
Vanadium	0.001 - 0.01	0.01 - 0.1

(continued)

TABLE 9: (continued)

Element	Stack Effluent, % Ashed Material	Collector Catch, % Ashed Material
Iron	0.1 - 5.0	0.5 - 5.0
Tin	0.001 - 0.5	0.05 - 0.5
Aluminum	0.1 - 10	1 - 10
Zinc	1 - 10	1 - 10
Magnesium	1 - 10	1 - 10
Titanium	0.5 - 5.0	0.5 - 5.0
Silver	0.0001 - 0.01	0.001 - 0.1
Boron	0.01 - 0.1	0.01 - 0.1
Barium	0.1 - 1.0	0.1 - 1.0
Beryllium	0.001 - 0.01	0.001 - 0.01
Calcium	10+ -	10+ -
Sodium	1 - 10	1 -
Lead	0.01 - 0.5	0.1 - 1.0
Sulfur	-	0.620 -
Phosphorus	1.140 - 1.460	1.760 -
Silicate	5.4 -	-

Source: PB 200,514 June 1971

GASEOUS EMISSIONS

Gaseous incinerator emissions are not, at least at the present, of primary concern as a source of air pollution. When compared to other gaseous emission sources, the contribution of incinerators is relatively small. It is for this reason that incinerator air pollution control equipment is adapted to the removal of particulate matter rather than gases. Some published data show that the wet collecting devices can remove small amounts of gases. The number of measurements of gaseous emissions, although not plentiful, is sufficient to provide some idea of the types and amounts of such emissions emitted from a municipal incinerator. Oxides of nitrogen, oxides of sulfur, aldehydes, hydrocarbons, and ammonia are emissions that have been detected and have been discussed in the literature.

Oxides of Nitrogen

Both batch and continuous-feed incinerators emit nitrogen oxide (NO) in small amounts that are not significantly different for a given amount of charged refuse. Actual tests indicated amounts ranging from 1.4 to 3.3 pounds per ton of refuse charged for a 50 ton per day batch-feed incinerator and a 250 ton per day continuous-feed incinerator.

Nitrogen dioxide (NO_2) emissions increase with increasing amounts of excess air as shown in Figure 4.1. Wet cleaning processes tend to increase slightly the amount of NO_2 produced. The amount of underfire air has a significant effect on NO_2 production (see Figure 4.2), an occurrence that has been explained as

FIGURE 4.1: RELATIONSHIP BETWEEN OXIDES OF NITROGEN AND EXCESS AIR IN 50-TON-PER-DAY UNITS

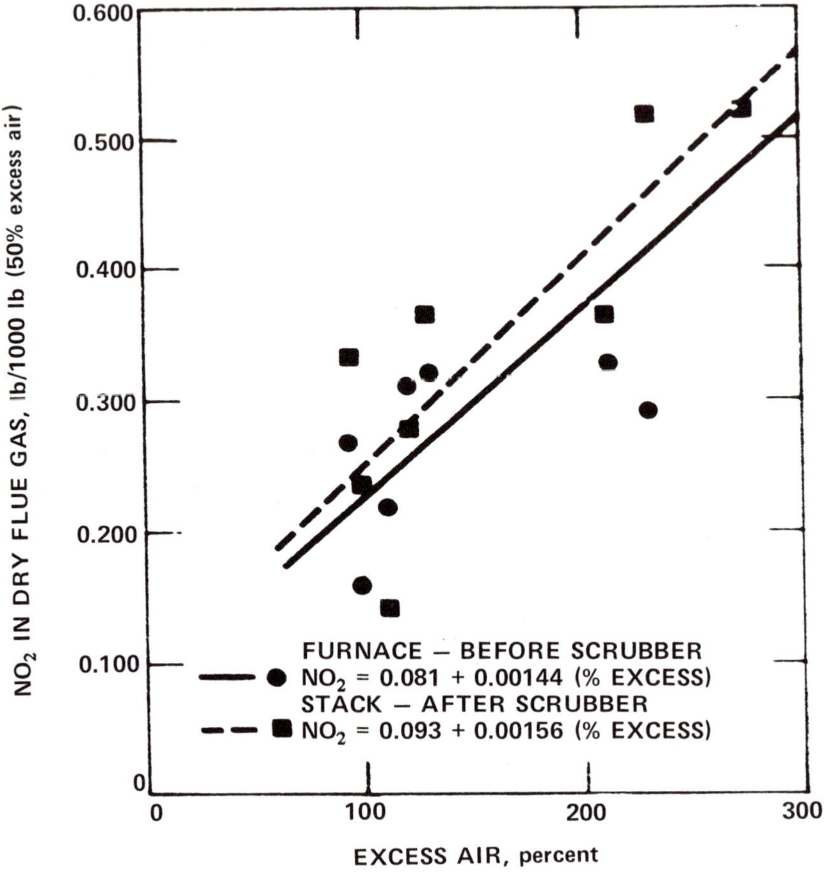

Source: PB 200,514 June 1971

being a result of the variance in oxygen consumption and its residence time with underfire air.

Variables found to be significant in the formation of nitrogen oxides, in the study by Rose were excess air and feed rate. There were no distinguishable effects on NO_2 produced by (1) pounds of fuel per charge, (2) stoking interval, (3) percent secondary air, and (4) percent underfire air. The main effect attributable to excess combustion air was a 12% increase in lb NO_2/hr and a 12% increase in lb NO_2/lb fuel burned. The main effect attributable to feed rate was a 44% increase in lb NO_2/hr and a 12% increase in lb NO_2/lb fuel burned.

FIGURE 4.2: RELATIONSHIP BETWEEN OXIDES OF NITROGEN AND UNDERFIRE AIR IN 250-TON-PER-DAY UNIT

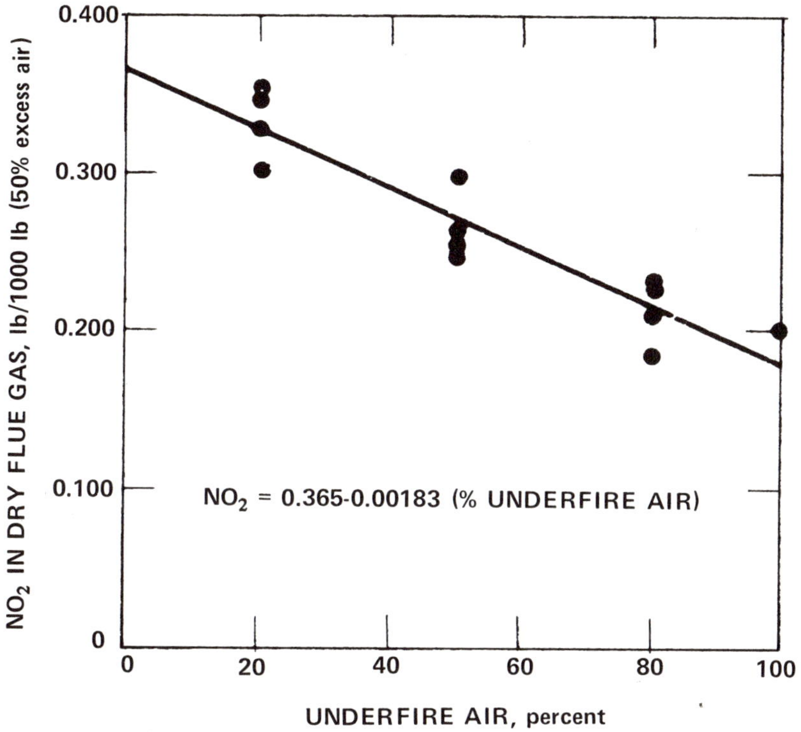

Source: PB 200,514 June 1971

Under all conditions tested, high instantaneous concentrations of nitrogen oxides in the flue gas occurred when gas temperatures at the bridge wall were high. As gas temperature at the bridge wall decreased, nitrogen oxides concentrations generally decreased.

In addition to fuel charging, agitation of the burning fuel bed was also responsible for increased burning as revealed by high gas temperatures at the bridge wall and correspondingly high nitrogen oxides concentrations. The critical temperature associated with nitrogen oxides formation is thought to be in or directly over the fuel bed, but gas temperature at the bridge wall may be a usable indicator of nitrogen oxides concentration in the effluent. Before establishing the relationship between gas temperature at the bridge wall and nitrogen oxides concentration in the effluent, the reproducibility of burning conditions as reflected by these parameters must be firmly established. It is possible that experimental variables found to be significantly related to oxides of nitrogen emission are effective only through their influence on temperature.

The most significant findings of the study by Rose, et al which can be applied to municipal and industrial incineration are the factors affecting the formation of oxides of nitrogen. Three factors found to control this formation were (1) temperature, (2) excess air, and (3) feed rate (grate loading).

The increase in oxides of nitrogen concentration was not linear with temperature but increased as a logarithmic function of temperature. While bridge wall gas temperature was used in establishing this relationship, it is realized that ignition zone temperatures are controlling. Of import to incinerator design, however, is the fact that increased operating temperatures result in increased oxides of nitrogen formation, other conditions being equal.

Increasing the excess combustion air resulted in increased nitrogen oxides formation for a given grate loading, despite the accompanying reduction in temperature. The effect of decrease in temperature with its corresponding decrease in oxides of nitrogen is apparently a less significant shift than the increase in oxides of nitrogen due to higher excess combustion air. This relationship should hold for larger incinerators wherein the temperature decrease will be less for a corresponding increase in excess air since the surface to volume ratio, for heat loss, will also be smaller. From an operational standpoint this effect is significant in that practices of reducing furnace temperatures through the use of high levels of excess air may result in increased total discharge of oxides of nitrogen.

As in the case of excess air, an increase in feed rate resulted in an increase in oxides of nitrogen formation. The cause of this effect, however, is obscured since the ignition chamber temperature also increased with increased feed rate at a given excess air level and the effect may be a result of temperature increase. This effect from the operational and design standpoint, however, is significant in that increased grate loadings can be expected to result in increased oxides of nitrogen emissions.

Carbon Dioxide

Stack emissions of carbon dioxide (CO_2) usually amount to only 1 to 6% of the dry volume. At the furnace exit, however, the amount of CO_2 is larger, ranging from approximately 4 to 16%. Carbon dioxide content in these concentrations is of little, if any, interest other than the information it can provide on the efficiency and rate of combustion of the refuse.

Carbon Monoxide

The concentration of carbon monoxide (CO) in gaseous incinerator effluents is so small that it is nearly impossible to detect. One of the largest recorded concentrations of carbon monoxide (1.0 lb/1,000 lb of dry flue gas) was at a 50-ton-per-day batch-feed incinerator during a firing that produced a low operating temperature. Carbon monoxide emitted from a 250-ton-per-day continuous-feed incinerator ranged from 0.03 to 0.07 pound per 1,000 pounds of dry flue gas. Emission of CO is a function of proper employment of the combustion air as it relates to (1) distribution within the ignition chamber and (2) levels of oxygen available in the combustion zone.

Oxides of Sulfur

Sulfur oxide emissions from municipal incinerators are practically negligible because the sulfur content in refuse generally averages only about 0.1%. For comparison, the sulfur content of coals fired in power plants averages approximately 1.0 to 2.5%. If the stack gas concentration of sulfur dioxide (SO_2) is measured in parts per million by volume, the amount of diluting excess air supplied to the furnace will directly affect the concentration of the SO_2. Published data of SO_2 emissions from municipal incinerators range from 0 to 100 parts per million by weight.

Figure 4.3 is a plot of SO_2 versus excess air data for incinerators in California and New York. The solid line gives the relationship between the SO_2 emission and excess air assuming the refuse contains 0.1% sulfur and conversion of sulfur into SO_2 is complete. The data suggest, however, that only a fraction of the sulfur is converted into SO_2, assuming the 0.1% sulfur content is approximately correct. The remaining sulfur can be contained in the residue, as has been confirmed by analyses.

Incinerators equipped with auxiliary burners for either waste-heat-recovery boilers or for burning low-calorific-value refuse often use high-sulfur-content fuels such as coal and oil. In this type of incinerator, SO_2 emissions are considerably higher and can be of considerable concern.

Formaldehyde

Formaldehyde is generated in municipal incinerators in minute quantities. The amount generated has been shown to be related to the temperature of the furnace gases, which in turn is related to the amount of excess air supplied and the amount of underfire air. For a 50-ton-per-day batch-feed incinerator operating at 108% excess air, no formaldehyde was produced. However, when the excess air was increased, which decreased furnace temperature by 400°F, up to 0.021 pound of formaldehyde per ton of charged refuse was formed. For a 250-ton-per-day continuous-feed incinerator operating at 185% excess air, 0.0014 pound of formaldehyde per ton of refuse was formed.

Hydrocarbon

Hydrocarbon content of incinerator flue gases is usually well below the limit of detectability of the measuring instruments. Measurement of hydrocarbon emissions from a 50-ton-per-day batch-feed incinerator shows that less than 0.003 pound per 1,000 pounds of dry flue gas is emitted when 110% excess

FIGURE 4.3: SO_2 CONCENTRATION IN MUNICIPAL INCINERATOR FLUE GASES

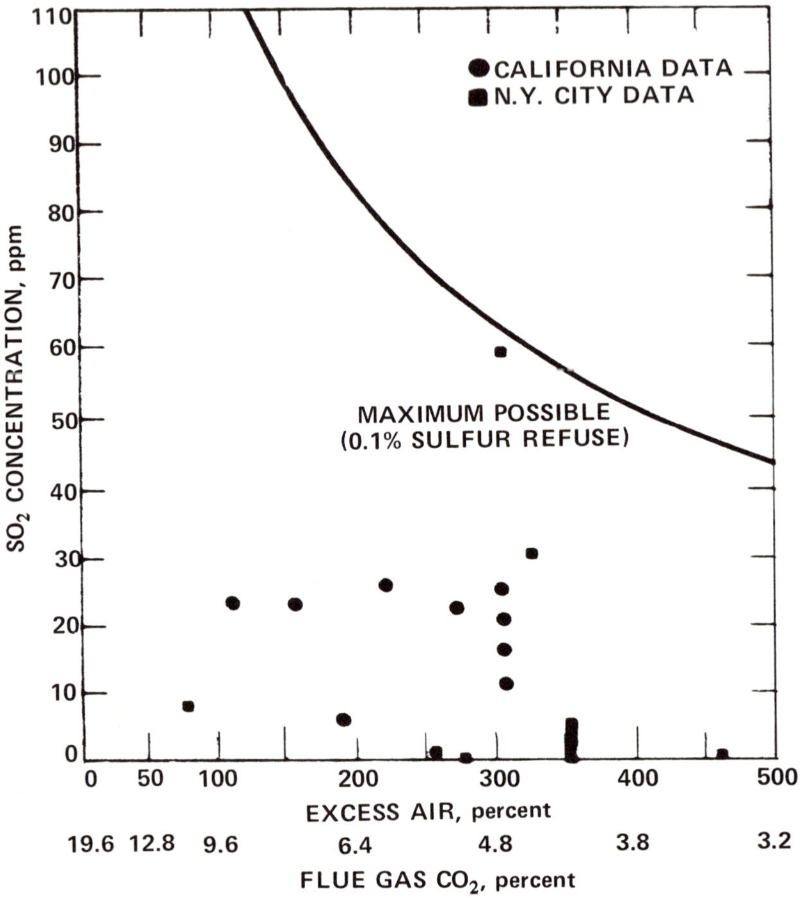

Source: PB 200,514 June 1971

air is used. For a 250-ton-per-day continuous-feed incinerator using 150 to 190% excess air, less than 0.08 pound of dry flue gas was measured. Polynuclear hydrocarbons have been detected in incinerator flue gases.

In the study by Rose et al, combustion of fuel at 50% excess combustion air produced measurable concentrations of hydrocarbons under all conditions of operation. Maximum concentrations ranging up to 400 ppm were produced with 50% excess air distributed between overfire and underfire. All conditions of combustion at 150% excess air, both with and without secondary air, failed to produce hydrocarbons in excess of the 30 ppm lower limit of detectability of the measuring system. At this level of excess air there was no significant difference in the total amount of hydrocarbons produced by varying (1) the fuel charging rate, (2) the stoking interval, (3) the distribution of combustion air, or (4) the lb of fuel per charge.

Chlorine

Chlorine has been found to be present in incinerator stack gases in rather minute quantities in the form of hydrogen chloride (HCl). Its concentration is dependent on the proportion of plastics (polyvinyl chloride) in the refuse which proportion in turn may be influenced rather strongly by the amount of industrial refuse processed by an incinerator. Maximum HCl content in European municipal incinerators is about 0.02 volume percent in wet flue gases. An increase in use of polyvinyl chlorides would make this emission become of more concern in municipal incinerator operations.

RESIDUE

The characteristics of the residue can determine the means and location of its ultimate disposal. The residue should be of a nature, or disposed of in such a manner, that insect and rodent attraction, dust, odor, and water pollution from leaching are at a minimum.

Residue consists of siftings that fall through the grates, as well as the "burned-out" refuse that remains at the end of the burning grate. The amount of residue that sifts through the grates is obviously dependent on the design and type of the grate. Some grates such as the rocking grate are designed to achieve as much sifting through the grate as is possible. Other grates are designed only with underfire air supply in mind and have much lower sifting rates. A study performed on two 200-ton-per-day rocking-grate stokers showed that 177 pounds of siftings was produced per ton of refuse burned. Table 10 summarizes the sifting weight and percentages for 325,100 pounds of fired refuse.

TABLE 10: SIFTING WEIGHTS AND PERCENTAGES

Grate	Sifting Weight, lb	% of Refuse
Feeder	1,160	0.36
Burner	27,600	8.49
Total	28,760	8.85

The composition of siftings varies with the position along the grate at which the sifting took place. Less ash, clinker, glass, ceramic material, stone, and metal sift through the feeder grate than sift through the burner grate. Feeder grate siftings contain more combustible material, moisture, and organic matter than burning-grate siftings.

Residue Composition

The physical analysis of total residue from a 300-ton-per-day continuous-feed incinerator is presented in Table 11. The data presented in this table are not meant to represent average or typical residue compositions. Rather, they identify the residue constituents for one specific municipal incinerator. They can be useful as a "first guess" in estimating residue composition. Table 12 presents typical ranges of values for the various residual constituents.

TABLE 11: CLASSIFICATION OF INCINERATOR RESIDUE

	Dry Weight, lb	Percent of Total
Ferrous metal	943.97	15.75
Tin cans	(828.25)	(13.82)
Other	(115.72)	(1.93)
Magnetic flakes	227.32	3.80
Nonferrous metal	18.04	0.30
Glass over ¼ inch	567.72	9.48
Ceramics, stones	90.53	1.51
Clinker over 3 inch	487.76	8.15
¼ x 3 inch	956.50	15.96
Ash, nonmagnetic		
¼ x 10 mesh	373.23	6.23
Minus 10 mesh	590,79	9.87
Combustible		
Paper, wood, char	107.43	1.79
Putrescible (visual)	4.11	0.07
Bones, pits	1.82	0.03
In conveyor water	1,620.90	27.06
Total	5,990.12	100.00
Less salt	-26.5	
Net residue	5,963.6	
Combustible-free residue	5,209.1	

TABLE 12: RESIDUE COMPOSITION (Percent)

Material	Range
Metals	19 - 30
Glass	9 - 44
Ceramics, stones	1 - 5
Clinkers	17 - 24
Ash*	14 - 16
Organic	1.5 - 9

*Exclusive of other materials listed.

Source: PB 200,514 June 1971

Potential water pollution from residue landfill sites by leaching can be a major consideration in selecting a location for a site. From 4.75 to 5.75% (by dry weight) of the residue is water soluble. Again, the variation can be the result of a complexity of factors such as incinerator design, incinerator operation, and refuse heterogeneity. An analysis of the water-soluble portion of residue for a batch- and continuous-feed incinerator is presented in Table 13.

TABLE 13: AVERAGE ANALYSIS OF WATER-SOLUBLE PORTION OF RESIDUE*

	Batch-Feed Incinerator	Continuous-Feed Incinerator
Hydrocarbon concentration	6.1666	9.1666
Alkalinity	0.1156	0.1865
Nitrate nitrogen x 10^{-4}	4.0078	3.48
Phosphate x 10^{-4}	2.75	4.416
Chloride	0.1221	0.0771
Sulfate	0.0813	0.2447
Sodium	0.04675	0.197
Potassium	0.04230	0.048
Iron	0.00617	0.015

*Percent by dry weight of sample.

Source: PB 200,514 June 1971

The residue composition for incinerators A-G previously described (p. 52) is detailed below in Table 14. It is expressed on a percent by weight "as-sampled" basis. The fines are defined as the unidentifiable materials passing through a 0.5-inch wire-mesh screen. The unburned combustibles are those visually identifiable combustible materials that pass through the incinerator without being burned. The unburned combustibles describe the visual appearance of the residue rather than the combustible content of the residue. The volatiles and heat content are more reliable indicators of combustible content. As stated previously, the fines also contain ferrous and nonferrous metals that are not determined during the separation procedure.

TABLE 14: RESIDUE COMPOSITION (Percent by Weight)

	Incinerator						
Component	**A***	**B***	**C***	**D**	**E**	**F**	**G**
Fines	44.9	52.5	38.9	36.4	74.5	79.4	52.6
Unburned combustibles	(**)	(**)	1.3	35.8	0.1	0.7	1.1
Metal	23.9	14.6	13.0	14.5	21.4	16.8	20.0
Glass, rock	31.2	32.9	46.8	13.3	4.0	3.1	26.3

*Dry samples
**Unburned combustibles included with fines.

Source: PB 216,372 (1970)

Incinerators E and F, rotary kilns, probably produced a higher percentage of fines than the other incinerators (74.5 and 79.4%, respectively) because the tumbling action of the kiln reduced the size of glass and rocks. The larger percent produced by Incinerator F, although not great, could be because its kiln is longer than that of Incinerator E, 30 ft compared with 23.

Incinerators C and D, a conical burner and a traveling grate, produced the lowest percentage of fines, 38.9 and 36.4%, respectively. The residue from Incinerator C fused into a slag that minimized the quantity of fines.

The residue from Plant D contained 35.8% unburned combustibles. The facility was overloaded during the study week because the refractories in one furnace had collapsed. Although the actual burning rate could not be determined because the solid waste was stockpiled during the time and no reliable estimate could be made of the quantities burned, waste was being processed as fast as possible. The obviously overloaded furnace and the lack of agitation on the traveling grate contributed to the high percentage of unburned combustibles.

EFFLUENT WATER

Water is used in various parts of the incineration process and, as it is used, it usually becomes contaminated with dissolved and suspended matter. The contamination of the waste water is generally great enough that the water must be treated prior to discharge to prevent or control pollution of rivers and underground water streams. Characteristics of waste water from five types of incinerators are given in Table 15. For comparison purposes some average characteristics of river water and sewage are also included in the table.

TABLE 15: CHARACTERISTICS OF INCINERATOR WASTE WATER

Source	Odor[a]	pH	Alkalinity, ppm	Total solids, ppm	Total volatile suspended solids, ppm	Percent	5-day biochemical oxygen demand at 20°C, ppm
River	0	7.0	50	100	10	trace	2
Sewage	4M	6.8	100	500	300	50	200
Incinerator wastes							
Batch-fed incinerator (rectangular)							
Ash hopper	3M			1,327	69	70	700
Fly ash disposal	1M	7.2	50	11,846	11	31	3.2
Lagoon	2S	7.3	134	9,580	13	24	54
Continuous-feed traveling grate							
Residue conveyor	2M	11.6	424	1,830	236	71	618
Batch-fed (circular)							
Residue conveyor	3M	4.6	330	6,302	56	78	750
Continuous-feed rocking grates							
Residue conveyor	3M	–	–	–	45	47	560
Continuous-feed reciprocating grate							
Residue conveyor	1S	6.4	410	–	14	43	605

[a]M - Moldy; S - Sulfuretted; Scale of 1 (no odor) to 5 (very strong).

Source: PB 200,514 June 1971

Reported studies of scrubber effluent waters from incinerators in Jacksonville and Ft. Lauderdale, Florida for a batch-feed incinerator, indicate that both the effluent water temperature and acidity varied widely. The pH ranged from 3 to 5 and the water temperature fluctuated from 140° to 180°F. Maximum values of pH were noted when the temperature of the effluent water was at or near a minimum. Table 16 summarizes the contributions of various chemical constituents made by a scrubber system in the Ft. Lauderdale incinerator. Cyanide and phenols show the largest increase, with rather substantial increases in many of the other constituents.

TABLE 16: ANALYSES OF SCRUBBER WATER AT FT. LAUDERDALE, FLORIDA, INCINERATOR (JUNE 1966)

Chemical Constituent	Raw Water	Scrubber Effluent	Contribution from Incineration
Iron (Fe) (mg/l)	0.35	2.00	1.65
Barium (Ba) (mg/l)	0.0	5.0	5.0
Cyanide (CN) (μg/l)	210.0	5,400.0	5,190.0
Chromium (Cr) (mg/l)	0.0	0.13	0.13
Lead (Pb) (mg/l)	0.0	1.30	1.30
Phenols (μg/l)	5.0	1,726.0	1,721.0
Copper (Cu) (mg/l)	0.08	0.18	0.10
Zinc (Zn) (mg/l)	0.0	2.40	2.40
Manganese (Mn) (mg/l)	0.0	0.30	0.30
Aluminum (Al) (mg/l)	0.18	20.80	20.62

Source: PB 200,514 June 1971

Refuse Incinerator Design

The material in this chapter is intended to provide the reader with exposure to considerations involved in refuse incineration design. Incinerator Institute of America's guidelines and other related engineering references should be consulted before actually initiating an incinerator design.

THE INCINERATOR DESIGN PROBLEM

Potentially, incineration should offer the most significant volume reduction of all solid waste disposal methods. According to a U.S. Bureau of Mines report, poor incineration will reduce the volume of refuse by 75 percent, while efficient burning can reduce it by 95 percent.

Inefficient incinerator design and operation mitigates against incinerators being a suitable solution for the urban solid waste problem. However, good incinerator design and the development and implementation of reliable control instruments and procedures present formidable problems as the processes occurring in an incinerator are very complex, as exemplified by the varied nature of the material being burnt. It is not surprising, therefore, to find that designers have had to rely on experience and empirical relationships rather than quantitative descriptions of the chemical kinetics and heat, mass and momentum transfer processes.

Typically, designs are based on gross overall heat and material balances, rule-of-thumb guidelines on the achievable burning rates per unit area of grate surface with different refuse types, and on allowable combustion intensities in the overfire volume (Table 1). Of these three requirements the most difficult to estimate has been the achievable burning rate. The heat and mass balances are straightforward to calculate, particularly as the air requirements for most solid fuels are remarkably uniform when expressed on a basis of energy liberated. The achievable burning rates per unit of grate are estimated from guidelines such as those provided by Table 1 or from the rule of thumb that the heat release rate within the fuel bed should be about 300,000 Btu/hr/ft^2 of grate area.

TABLE 1: PARAMETERS OF DESIGN FOR REFUSE INCINERATORS

	Type of Refuse*	Grate Loadings in Lbs. of Refuse per Hour of Operation per square foot of Grate Area	Volume in Cubic Feet per Ton of Refuse per 24 Hours	
			Furnace (Primary) Chamber	Combustion (Secondary) Chamber
Range of Values	M	58 to 109	8.5 to 25.0	12.1 to 28.0
	R	50 to 72	13.4 to 14.5	26.6 to 31.8
	C	54 to 98	9.9 to 13.8	17.2 to 28.3
Average Values	M	77	12.7	18.5
	R	58	13.6	29.9
	C	77	11.5	21.3

*M - Mixed refuse made up of garbage, rubbish, and noncombustibles.
R - Refuse comprised of burnable rubbish only.
C - Refuse containing combustibles only, such as garbage and burnable rubbish.

Source: PB 223,626 September 1973

For a typical as-fired heating value for refuse (5,000 Btu/lb), the burning rate for the latter criterion would be 60 lb/hr/ft^2 of grate area which agrees with the Incinerator Institute of America's guideline of 60 to 65 lb/hr/ft^2 for a Class V incinerator and the values given in Table 1. The establishment of these guidelines has ignored the fundamental problem in many engineering operations, namely the calculation of the rate of the process, and has perhaps established a precedent of perpetuating inefficient incinerator designs.

The maximum allowable heat release rate within a fuel bed should be determined from consideration of the maximum bed temperature that would prevail and therefore must take heat losses from the bed into account; temperatures that are too high cause difficulties with clinkering and problems with grates clogged with molten glass and aluminum. There are indications that successful operation has been achieved at heat release rates up to three times the suggested maximum value of 300,000 Btu/hr/ft^2, hinting at the tempting prospect of reduced investment cost per ton of refuse processed.

A similar situation to that encountered when selecting grate sizes is found in the overfire region, where few design criteria are available. The only specifications given are for furnace volumes based on guidelines such as those given in Table 1, or on the rule of thumb that the volumetric heat release should be around 20,000 Btu/hr/ft^2, and on allowable gas velocities at different points in the incinerator.

The value of the maximum combustion intensity is given without any regard for the amount of combustible that has to be burnt in the overfire region. No guidelines are given as to the desirable ratio of primary air introduced through the fuel bed to secondary air injected into the overfire region. There is general agreement in the literature that the overfire air must be supplied with sufficient momentum to provide adequate mixing with the combustible gases. There are, however, no reliable methods presently available for determining how this should be accomplished.

Designs based on these methods may have been satisfactory in the past, but, with the effects of the new legislation concerning acceptable levels of gaseous and particulate emissions as well as ash and residue quality beginning to be felt, increasing numbers of practitioners are becoming interested in developing more fundamental ways of designing their incinerators.

For example, there is a need for indicators for determining the probable burning characteristics of different fuels, and the selection of the size, number and location of overfire jet systems and the ratios of primary and secondary air. In addition, long-term trends in refuse quality have suggested that the volatile content of refuse, which is a measure of quantity of overfire air that is required to complete the combustion of the volatile products distilling from the fuel bed, will increase over the years (Table 2).

The projected increase of this component in refuse is also expected to require that more significance be placed on the successful operation of the overfire air jets. This will require a more sophisticated approach to the design of these jets.

The problems of operating an incinerator effectively are enormous when one realizes the tremendous variation in feed material that is handled from day to day.

The variation in the feedstock quality of an incinerator is orders of magnitude greater than that in a pulverized coal fired utility boiler (or its chain grate stoker predecessor), yet incinerator controls are barely existent compared to the sophisticated controls of a modern utility boiler.

TABLE 2: GENERATED REFUSE COMPOSITION—HEATING VALUE AND QUANTITY*

	1970	1975	1980	1990	2000
Composition					
(weight %, as discarded)					
Paper	37.4	39.2	40.1	43.4	48.0
Yard Wastes	13.9	13.3	12.9	12.3	11.9
Food Wastes	20.0	17.8	16.1	14.0	12.1
Glass	9.0	9.9	10.2	9.5	8.1
Metal	8.4	8.6	8.9	8.6	7.1
Wood	3.1	2.7	2.4	2.0	1.6
Textiles	2.2	2.3	2.3	2.7	3.1
Leather and Rubber	1.2	1.2	1.2	1.2	1.3
Plastics	1.4	2.1	3.0	3.9	4.7
Miscellaneous	3.4	3.0	2.7	2.4	2.1
(weight %, as burned)					
Moisture	25.1	23.3	22.0	20.5	19.9
Volatile Carbon	19.6	20.1	20.6	21.8	23.4
Total Ash	22.7	23.4	23.9	22.8	20.1
Ash (excluding glass and metal)	6.5	6.2	6.1	6.0	6.0

*Projected average 1970-2000

Source: PB 223,626 September 1973

SITE SELECTION FOR REFUSE INCINERATORS

When incineration has been chosen as the method of refuse disposal, the selection of a proper site for the incinerator plant is important. Since conditions in different localities are variable, it is difficult to specify site requirements for all. Economy and appearance are important considerations. If possible, an inconspicuous location central to collection areas and easily accessible by road is desirable.

Municipal incinerators are usually located as near as feasible to the population centers served, except for instances in which bias is sometimes given in the direction of the maximum projected population growth. Convenient road access to the incinerator from all areas to be served is important. If a decision must be made between a "close-in" site or a more remote site, hauling distances can be a major consideration.

Normally, residue is more economical to haul than raw refuse, if great distances are involved. Industrial areas are usually more receptive to incinerators than are nonindustrial areas. Purchase of a generous amount of acreage assures room for future expansion and is usually a good investment. Locating incinerators near sewage plants for sludge treatment or incineration of odorous gases emitted during sewage treatment can have its merits.

The distribution of incinerators, by community size as of 1965, is given in Table 3. Cities with a population of 1 million or more have the lowest percentage of incinerators, while cities with populations of 10,000 to 24,900 have the highest, i.e., 25.2 percent.

TABLE 3

Community Population, Thousands	Number of Communities in US, 1960	Percentage of Communities with Incinerators	Average Number of Incinerators per Community	Distribution of Incinerators by Community Size Number	Distribution of Incinerators by Community Size Percent
1,000 or over	5	80.0	4	16	5.1
500 to 999.9	16	75.0	2	24	7.6
250 to 499.9	30	50.0	1.5	22	7.0
100 to 249.9	81	30.0	1	24	7.6
50 to 99.9	201	25.0	1	50	15.9

Source: PB 200,514 June 1971

A ramp is necessary for reaching the charging floor of a direct charge plant, therefore, a side hill site would be worth looking for to reduce construction cost.

Care should be taken to avoid a location where a down draft of the stack effluent is caused by the prevailing wind. A bank of high trees in the vicinity may cause stack discharge depressions; or the same effect might be had from the proximity of a hill or ravine.

If this information is not available through knowledge of local conditions, air currents may be traced by the use of balloons to determine whether certain winds will, or will not, bring the stack discharge down. The fumes issuing from an incinerator chimney, even if the plant is operating efficiently, will be noticeable and, while perhaps not injurious to health, are undesirable under the above down draft conditions.

BASIC INCINERATOR COMPONENT DESIGNS

Trucks deliver refuse to a storage pit at most modern municipal incinerators. The size of the storage pit at such an incinerator is dependent on such factors as capacity of the furnace, emergency storage required in the event of furnace breakdowns, and refuse truck pickup schedules.

The unloading facilities needed at the plant will depend on the type of vehicles employed and whether or not the refuse is to be stored or dumped directly into the incinerator. Loading platforms should be large enough to facilitate charging of material.

Refuse trucks can be unloaded directly onto a charging floor. When this is done, mounds of refuse are heaped in open storage in order to supply the furnaces continuously and to minimize the waiting and unloading time for the trucks. Many incinerators have wing walls, barricades, and charging doors for the introduction of refuse into the primary furnace chamber. Minimum unloading expense is obtained by having the refuse truck unload directly at such wing gate inlets, so that little labor is required to introduce the refuse material.

Similarly, it is possible to employ small hoppers directly above the inlet of the furnace into which the refuse truck may unload. Direct charging is common in plants which operate for only 8 to 12 hours per day. The collected material is burned immediately with the exception of small quantities of refuse temporarily left in open storage. Hand charging, due to labor costs, is generally the most expensive of all methods.

Another method of handling refuse involves the discharge of segregated or semisegregated material onto a sorting floor for the salvage of materials. Salvage is governed mainly by the local market for salvage material.

In large plants, it is necessary to have storage bins with enough capacity to allow continuous plant operation around the clock. Bin design criteria generally requires a 16 to 24 hour reserve of refuse to be burned. The bins should have pitched floors and drains to facilitate cleaning. In units with large storage bins, traveling bridge cranes or monorails are used for charging the primary furnace.

Refuse is introduced into the primary chamber of an incinerator through a charging door. This opening can be either on the side of the furnace or on top. Most plants which have been constructed recently have employed top charging because of operating practicalities. Air pollution control authorities, however, advocate side charging because it tends toward more uniform feeding, which enhances optimum combustion. The convenience of top charging is a desirable feature. Side charging should be seriously considered where air pollution aspects are critical. In either case, it is necessary that overloading be prevented to avoid incomplete combustion.

The dimensions of the charging doors vary. Normally they are about two feet square or two feet in diameter. They are fixed by considering overfire-air requirements, the charging method employed, and/or the type of refuse burned. The charging doors are usually opened only when the refuse is being introduced. However, they may be left open to provide an additional source of overfire-air to meet the demands of wet material. Door stops should be provided for this purpose. In a plant where there is to be a minimum of hand feeding, and a mechanical conveyor or a bulldozer is used to introduce the charge, larger openings are desirable. As for the type of refuse burned, large boxes or other bulky refuse should not govern the size of charging doors, unless a great amount of this type of refuse makes it more efficient to introduce this material without first breaking it up.

The main consideration, however, should be ease in opening the doors and controlling excess air. The charging doors are generally lined with firebrick to enable them to withstand the high temperatures generated in the primary chamber. Counterweighted guillotine-type doors are useful on units that are charged from the side.

Stoking doors are used on furnaces even when top charging is employed. These inlets are used for stoking, introducing dead animals, controlling the amount of air, and for observing furnace operation. These side openings are generally covered with hinged doors. Stoking doors and cleanout doors should be provided with adjustable air ports for additional control of excess air. On larger units, counterweighted guillotine-type stoking doors are often used. Whatever the furnace opening, plant operators should be instructed to stand to one side when working as there is always the possibility of a dangerous explosion within the furnace.

Elevated cranes deliver the refuse to a charging hopper that feeds the refuse automatically through a chute to the feeder and drier stoker. The refuse is usually ignited on the feeder stoker before it is dumped onto the burner stoker. Air is supplied for combustion and temperature control through the grate, sidewalls, and roof of the combustion chamber.

Residue is discharged from the end of the stoker into mechanical conveyors that transfer the residue to storage bins or trucks. Residue is wetted occasionally to control dust. In some incinerators, combustion gases are passed into a second combustion chamber (secondary combustion chamber) to complete combustion of gases and entrained solids. Combustion gases are then cleaned prior to exhausting through the stack.

Scales

Many incinerators maintain accurate records of the amount of refuse processed. Weight is the usual record maintained and can be estimated by two methods. An accurate record is kept where a scale is installed, so that trucks can be weighed prior to discharging. If a scale is not available, the number of truck loads multiplied by the estimated weight per load will give an approximate figure. This method is not considered to be good practice, however,

Storage Pits

Several factors must be considered in the design of storage pits. As previously mentioned, furnace capacity, emergency storage, and truck pickup schedules are important factors to be considered in determining the size of the storage pit. Refuse can be dumped directly into the pit from the trucks or onto a conveyor belt that carries the refuse to the pit. Some charging hoppers are designed to receive truck loads directly, thereby eliminating the use of a storage pit except when the charging hoppers are full. An advantage of direct loading of the charging hoppers is that old refuse on the bottom of the hopper is burned first. Refuse would not build up in corners of storage pits where cranes cannot reach.

Trucks are almost exclusively the means by which refuse is delivered to municipal incinerators in the United States. In foreign countries, however, refuse is delivered to some plants by ships and railroad cars. So many additional variables exist in the design of receiving systems for ships and railroad cars that there is no one generalized design. An example of a custom design is a turntable and car dumper that empties refuse into the storage pit at the Stuttgart incinerator.

Dust generated during refuse dumping, crane loading, and hopper charging can be troublesome. Some plants furnish their employees with breathing masks if dust control methods are not used or are ineffective. Exhaust hoods over dumping areas can reduce the dust. Another method uses air inlet ports around the top and bottom of the pit. The upper ports draw in dust-laden air and the lower (near the bottom of the pit) drain off dangerous gases that occasionally form in the bottom of pits.

Cranes

Refuse is transferred from the storage pit to the charging hoppers by means of overhead, traveling cranes that can be equipped with either grapple or clamshell buckets. The refuse can be rearranged in the storage pits to permit truck dumping space. In the United States, grapples are more widely used than clamshell buckets are.

Clamshells do not have the grappling ability of grapples but are useful for cleaning the bottom of the pit. Buckets must have sufficient digging ability to pick up the refuse, and cranes should provide a means of preventing bucket twist and have a desirable operating speed. A steel grating in the floor can provide a wearproof parking place for buckets not in use.

In Europe both orange peel (polyp) and clamshell buckets are used, but polyp buckets are the more popular. Polyp buckets are more expensive, but have the ability to pick up different types of refuse more positively. Bucket capacities of European incinerators range from 5 to 7 cubic yards, which is larger than the 2 to 3 cubic yard capacity of buckets in the United States. European cranes work at slower operating cycles, which may cause them to use less power. This slower speed results in less damage to pit walls and hoppers, and makes a crane-weighing operation more practical.

At least two cranes are needed for the average refuse incinerator. The need for a third crane as a standby is stressed by some, debated by others, and denied by still others. Emergency crane repairs can be made in a short time by a well-trained crew with available spare parts.

There are two different types of crane installations. The bridge crane affords the most versatility by allowing movement in both directions over the storage pit and charging-hopper area. The second type, the more inexpensive monorail crane, moves only in one direction, that is, along the centerline of the bin. The width of the bin for this type of crane should not exceed by 2 or 3 feet the width of the bucket in its wide-open position.

The value of a good crane operator must not be overlooked. He can select the refuse from the pit to provide the most suitable mixture for incineration when the pit refuse is usually nonhomogeneous. He can remove large pieces of refuse that may not feed or burn satisfactorily. In some instances, the number of crane loads are counted or a crane is fitted with a scale to determine the amount of refuse fed to the incinerator.

Charging Hoppers and Gates

The charging hopper is the beginning of the completely mechanized portion of the incinerator. Hoppers of either metal or concrete are constructed in such a manner that they funnel the refuse by gravity through a chute to the furnace-charging mechanism. The flow of refuse can be shut off or regulated by a charging (hopper) gate (see Figure 5.1).

Hoppers can be fitted with eccentrically weighted rotors that make the hoppers oscillate, thereby controlling the flow of refuse to the furnace-charging mechanism. This method is particularly appropriate in ram-fed incinerators where the oscillating hopper can be synchronized with the furnace-charging cycle.

In continuous-feed furnaces the refuse in the charging hopper and chute seals off the heat of the furnace. To prevent fires in the charging hoppers the lower portions of the hoppers are connected to a water-cooled feeding chute through which the refuse passes to the charging gate.

Furnace Grates

The grates in a furnace are one of the most important parts of a continuous-feed incinerator. If refuse were merely dumped on a grate and burned without turning or agitation, burning would take place only on the top. Refuse not exposed to the flame and that next to the grate would leave the furnace incompletely burned.

Well-designed grates turn and agitate refuse as they move it through the furnace so that (1) a high percentage of the moisture is evaporated, (2) volatiles are gasified, (3) burnable solids are heated to ignition temperature, and (4) nonburnable refuse is heated to approximately 1500°F to make it nonputrescible.

Traveling Grates – Traveling grates, perhaps used more widely in the United States than elsewhere, provide movement of refuse through the furnace by means of continuous, conveyor-type movement. They are installed in line, usually in number of two or more.

The first section of a traveling-grate system is sometimes called the feeder grate. It is inclined and drops the refuse onto the burner grate (second section) to provide turning and agitation. Ignition of the refuse on the feeder grate normally takes place at about the middle of the grate. The speed of the feeder grate is controlled to provide sufficient drying and timely ignition of the refuse.

The burner grate (or grates) is horizontal; its speed is adjustable to fit the combustion nature of the refuse. The speed can be adjusted independently of the feeder grate.

One of the later developments in traveling grates has been the addition of more burner grates to provide additional dropoffs, and thus additional turnover to provide more complete combustion.

Reciprocating Grate – Another popular grate is the reciprocating grate, which advances and agitates the refuse by means of alternate rows of grates sliding back and forth over a stationary row of like grates.

The Von Roll System which is widely used in Europe, uses a reciprocating grate. Because improvements are constantly being made in each new installation, no one installation can be classified as typical. In a recent installation, the drying stoker is inclined 20 percent. There is a 5 foot drop from the drier stoker to the first burner stoker, which is inclined 30 percent. Another drop of 5 feet moves the refuse to the second burner grate, which is inclined 33 percent. Rogus states:

> The three stokers or grates are comprised of stepped-down special-cast steel pallets. These are alternately of solid and perforated bar key construction. The solid pallets have large swiveled arched inserts. The slow reciprocating downward movement, about a 5 to 6 inch stroke, of the individual pallets combined with the relative motion between them is augmented by the upward lifting action of the normally recessed segments.
>
> The overall effect provides a thorough intermixing and agitating action which promotes a near-complete burndown of the refuse. The siftings that pass through the grate system are discharged into zoned hoppers and then through gravity chutes into the underlying residue troughs.

Another European grate design, known as the Martin System, is a reverse reciprocating-grate system. This grate has a high efficiency, permitting the use of a single grate per furnace. The drying and burning is accomplished on a single short, but wide inclined grate. The stepped-down grate is sloped at an angle of approximately 30 degrees. The grate consists of heavy serrated cast-steel bars of chrome iron that can withstand temperatures up to 1600° to 1700°F. The grate consists of alternate reciprocating and stationary bars. The stoking bars actually push the refuse uphill against the downhill, gravity-induced movement of the refuse. This system has been used successfully in many countries with refuse of several types and widely varying calorific value and composition.

Rocker-Arm Grates – Rocker-arm stokers consist of row of grates that pivot up from the horizontal through an angle of 90 degrees and then back to the horizontal. The pivoting motion is alternated between odd and even numbered rows, which provides agitation and movement of the refuse through the furnace.

Barrel Grate (Drum Grate) – One of the more recent designs that is being used in Europe is the barrel-grate incinerator. In an incinerator in Dusseldorf, Germany, each furnace is equipped with seven contiguous cylinders set at progressively lower levels toward the discharge end at a slope of about 30 degrees from the horizontal.

In actuality, this design simulates a series of traveling grates, equal in length to the exposed perimeter of the barrel grate. The speed of the barrels is independently variable; the first grate rotates at a speed of 50 feet per hour and the last grate rotates at 15 feet per hour. The grates are 5 feet in diameter and 10 feet long. They are made of serrated cast iron arched segments that are keyed to a structural steel frame.

Rotary-Kiln Grate – Rotary kilns can be used for both drying and burning refuse. The refuse is constantly tumbled as it moves slowly under the action of gravity through inclined rotating kilns. Rotary kilns are used in combination with other types of grates, such as, where the refuse is dried and partially burned on a reciprocating grate that then delivers the burning refuse to a rotary kiln for final burning.

Batch Incinerator Design – There are many grate designs, particularly for batch-feed incinerators, which, although not discussed previously, should be mentioned. Batch-feed furnaces are usually equipped with one of five different grate designs: manually stoked, circular manually stoked, rocking cell, reciprocating, and oscillating.

Some generalizations concerning United States incinerators are in order. The concept of municipal incineration grew rapidly following World War II. Batch-feed incinerators were built more often than continuous-feed incinerators until 1963, after which the trend was reversed. The three most popular continuous-grate designs are the traveling grate, reciprocating grate, and the rocking grate. The most significant trends for batch-feed grates are the replacement of the hand-stoked grates with mechanically stoked reciprocating and rocking grates.

Grate-Burning Rate – The burning rate for grates is determined by the amount of refuse that can be burned per unit grate area per unit time and is commonly expressed as pounds of refuse per square foot per hour. Table 4 presents grate-burning rates for 157 municipal incinerators of various design in the United States. The year-to-year column gives the first and latest year of incinerator design for which a burning rate was reported.

TABLE 4: GRATE-BURNING RATES

Stoker Type	Feed Type	Year to Year	Number Reporting	Refuse Burned, lb/ft² grate surface-hour Max.	Min.	Median	Average
Manual	Batch	1946 - 1958	8	91	37.7	47/67	59
Circular	Batch	1945 - 1965	59	110	45.4	84	83.3
		1961 - 1965	2	70	70	-	70
Rocking	Batch	1949 - U.C.	37	71	32.4	57.5	56.8
		1961 - 1965	10	60	43	57/57.5	56.0
Rocking	*	1963 - U.C.	9	67.5	50	58	58.7
Traveling	*	1954 - U.C.	23	70	55.5	65	64.3
		1961 - U.C.	15	70	55.5	65	63.7
Reciprocating	Batch	1959 - 1965	11	87	35	57	57
		1961 - 1965	9	60	35	57	53.6
Reciprocating	*	1963 - U.C.	8	75	55.5	60	62.9
Oscillating	Batch	1958 - 1964	2	69.5	60	-	64.8
		1961 - 1964	1	-	-	69.5	-

*Continuous feed

Source: PB 200,514 June 1971

Combustion Chambers

There are basically two types of furnace wall construction, refractory and water-cooled structural steel, the choice of which can depend to a large extent on the sophistication of the gas-cleaning equipment used and on whether a large amount of waste heat is to be recovered. For these reasons, refractories have been used almost exclusively in the United States, while other countries with sophisticated gas-cleaning equipment and more emphasis on waste-heat recovery have made extensive use of structural steel (water walls). There are, of course, many other considerations in the choice between these two types of furnace wall construction.

Water-Walled Combustion Chamber – Combustion chambers of water-walled furnaces are normally lined with structural steel tubes through which water is circulated for the generation of steam. In most incinerators the tubes are welded together to form an integral wall. Water walls are heated almost entirely by radiation supplemented by some convection, and their presence has a tremendous cooling effect on furnace temperatures and substantially reduces the amount of excess air required for cooling the furnace. The use of small amounts of excess air has two major advantages. The first advantage is that the temperature of combustion of the refuse increases with decreasing amounts of excess air as illustrated in Figure 5.1. Obviously, the higher the temperature the more complete will be the combustion of the refuse. From Figure 5.1, the combustion temperature for 30 percent excess air is approximately 2500°F, which is substantially higher than that for refractory furnaces requiring large amount of excess air for cooling. A second advantage is that small amounts of excess air mean smaller amounts of gases that must be expensively cleaned in areas with strict air pollution control codes. The figure below shows gas temperature/excess air rate relationships for municipal refuse.

FIGURE 5.1: GAS TEMPERATURE VERSUS EXCESS AIR RATES

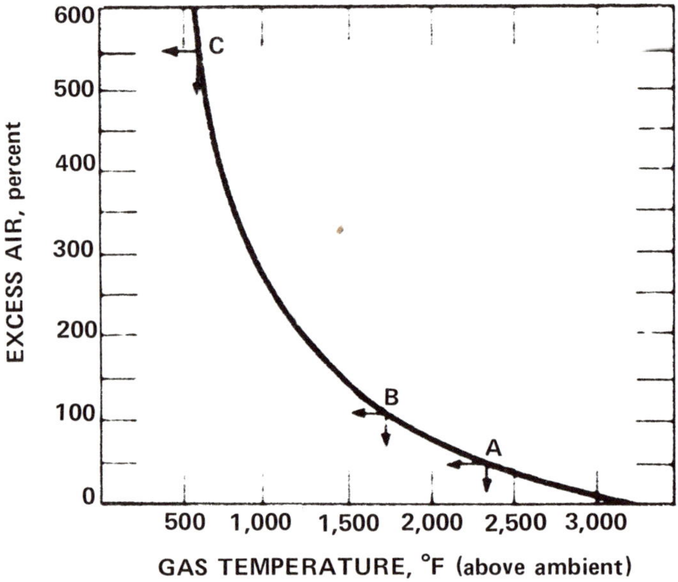

A = Gas temperature for 50 percent air = 2400°F
B = Excess air for furnace
Exit temperature = 1800°F = 110 percent
C = Excess air for gas temperature
At dust collector - 550 percent

Source: PB 200,514 June 1971

A cost comparison study of a water wall versus a refractory furnace for the Norfolk, Virginia, Naval Base installation showed that initial cost for the waterwall installation using 100 percent excess air was nearly equal to the initial cost of a refractory furnace using 200 percent excess air. Steam production from a given amount of refuse can be increased by approximately 38 percent with water walls. The use of water walls in other countries is extensive principally because of the requirement of low gas temperature for sophisticated gas-cleaning equipment (electrostatic precipitators) and the emphasis on waste-heat recovery. A relatively large water wall incinerator built at Issy-Les-Moulineaux, a suburb of Paris, uses four 17-ton-per-hour furnaces. Another noteworthy, water wall incinerator that will soon be placed into operation is at Ivry, another suburb of Paris.

Refractory Combustion Chamber – As with the water wall combustion chamber, the primary function of the refractory combustion chamber is to provide an enclosure where controlled combustion of refuse can take place. Because of the increasing size of today's municipal incinerators, refractory enclosures for large installations become quite large and sophisticated in design.

Widely fluctuating temperatures inside the combustion chamber, resulting for the most part from the varying calorific value of charged refuse, cause uneven expansions and contractions resulting in thermal shocks to the refractory lining. Measurements of the temperature variations of the refractory lining at various locations in the Oceanside Refuse Disposal Plant, New York, show that variations of several hundred degrees Fahrenheit do indeed occur in periods of less than 1 hour and that temperature differences from one location in the furnace to another frequently amount to several hundred degrees.

Refractories in common use are super-duty fireclay, high alumina, chrome magnesite, and plastics. Plastics are made from clays similar to those used in bricks. However, the plastics are prepared at the factory and shipped in a wet mix form. After the plastics are placed in the incinerator they are uniformly heated to a specified temperature during which time they develop into a ceramic-like structure and bond. For all refractories, uneven expansion and contraction can and must be appropriately accounted for. The refractory can be either self-supporting or be hung from a structural steel superstructure. Arches and walls supported by structural steel superstructures have proven successful in large incinerators.

Refractories can be constructed in sections so that the load for each section is independently carried through support castings to the superstructure, eliminating cumulative loading. Expansion joints for each section permit independent expansion and contraction eliminating the accumulation of thermal stresses. In the sectional design, refractory thickness is not required for wall stability and support; it is determined basically by temperatures and the operating conditions of the furnace. Thicker refractories are used when higher temperatures occur and when heat storage is necessary to control widely fluctuating temperatures. Refractory linings can be either air-cooled or insulated. Air-cooled walls can be used as ducts for delivery of over-fired air into the combustion chamber.

Incinerator Slag – The buildup of slag on the side walls of combustion chambers of refractory incinerators has become an increasing problem with the advent of continuous-feed incinerators. Long operating periods of several days at a time result in high wall temperatures that enhance slag buildup. Slag buildup is greatest on the lower, side walls of the furnace where it causes obstruction of the grates and burning refuse. The composition of slag varies over a wide range from one incinerator to another.

Several methods for reducing or preventing the formation of slag deposits on furnace walls have been used. Silicon carbide in conjunction with air cooling of the furnace walls has been used in some incinerators. The use of steam spray nozzles in a steam pipe mounted just above the grate has proven successful at the Oceanside Refuse Disposal Plant. Water walls in some of the European incinerators have alleviated this problem.

Heat Recovery Boilers

Boilers placed in the path of combustion gases can be both an effective and economical method of cooling the gases. For example, heat absorption by water-walled furnaces with well designed boilers can cause a gas temperature reduction of from 2500° to 450°F.

There are two basic boiler sections in the modern water wall, waste heat recovery incinerator. The first of these is the convection section, which is located immediately beyond the combustion chamber. In this section the gases move vertically upward passing through a series of boiler tubes. The velocity of the gases normally does not exceed 30 feet per second in this section.

Since there is still a large amount of entrained fly ash, and temperatures are high, the boiler tubes must be spaced far enough apart to prevent foul bridging across the tubes. On leaving the convection section, the gases have been reduced in temperatures to nearly 1000°F and are then usually channeled to move downward through the second boiler section, which is called the economizer. The tubes in the economizer are much more closely spaced because the fouling problem is reduced (flyash is less sticky) at the lower gas temperature. On leaving the economizer, the gases are ready for the gas-cleaning operation.

Auxiliary Heat

Auxiliary heat is sometimes used to attain high temperatures for the drying, ignition, and complete combustion of high moisture content refuse. Auxiliary burners may be installed in waste heat recovery incinerators to augment steam production on an as-needed basis when steam production from refuse drops below a specified amount.

Oil and gas, and sometimes coal, are used for fuel. No one location for the burner is universally accepted. It may be located directly in the incinerator furnace as illustrated in Figure 5.2, or installed in a separate combustion chamber, in which case the combustion gases from the burner and the incinerator come together at the boiler inlet.

FIGURE 5.2: AUXILIARY OIL BURNERS IN DUSSELDORF, GERMANY, INCINERATOR

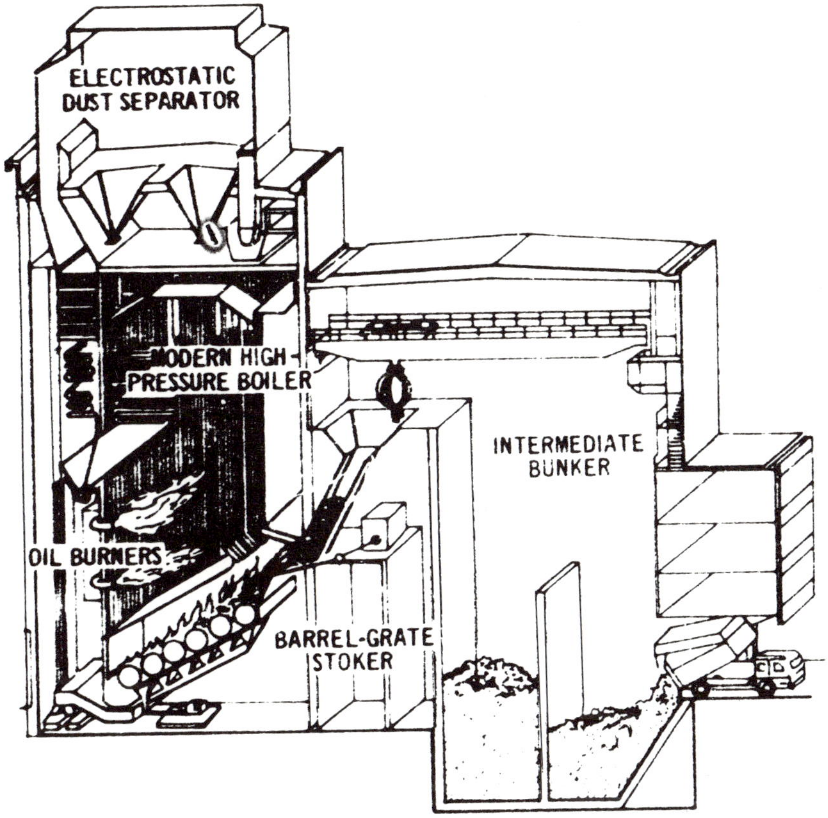

Source: PB 200,514 June 1971

Size Estimation for Incinerators

The size of an incinerator is governed by the amount of refuse to be consumed on a pound per day basis. It is very important that this calculation be made as accurately as possible. Direct quantitative measurements yield the best data.

A study must be made of the area from which the refuse is to be collected to determine the contributing population. Growth should not be overlooked. Therefore, it is advisable to consider the average peak load over a period of at least ten years hence. Extreme peak loads contributed by holiday crowds, should not be used in the data as excessive loads may be incinerated by providing for overtime use of the plant.

TABLE 5: BASIC REFUSE QUANTITY DATA

		Pounds per Capita per Day			Cubic Feet per Capita per Day			% of Garbage and Rubbish in Refuse	
Facility	Types of Waste	Average Minimum	Average Maximum	Median of Average	Average Minimum	Average Maximum	Median of Average	By Weight	By Volume
Lodges & hotels	Combined refuse	1.79	3.90	3.5	0.43	0.59	0.43	-	-
	Garbage	1.89	5.04	2.0	0.08	0.13	0.09	60	15
	Rubbish	1.15	3.24	1.3	0.33	0.67	0.37	40	85
Campgrounds	Combined refuse	0.45	1.60	1.4	0.07	0.41	0.18	-	-
	Garbage	-	-	-	-	-	-	40	20
	Rubbish	-	-	-	-	-	-	60	80
Picnic areas	Combined refuse	-	-	0.4	-	-	0.07	-	-
Park hdqrs. area (including residences and maintenance areas)	Combined refuse	1.52	2.82	2.0	0.28	0.51	0.40	-	-
	Garbage	-	-	-	-	-	-	53	20
	Rubbish	-	-	-	-	-	-	47	80

Source: PB 214,882 June 1958

Incinerators should be designed to consume the expected refuse during the expected charging time. The actual charging time should, where applicable, allow time for proper heating up, cooling and cleaning. For noncontinuous operations, automatic controls may be installed as optional equipment for the purpose of heating up the incinerator before the first load arrives in the morning, and thus, increasing the charging time by one to two hours.

Load estimates obtained during a 1954 field study of recreational facilities are presented in Table 5. Ashes are excluded from the table; the relative error is small. The figures shown are for facilities which are not necessarily in the same location. Therefore, amounts of garbage plus rubbish do not always equal the amount of combined refuse.

In areas which produce less than five tons of refuse per day, a single-unit plant (one furnace) is best. In such a small installation, no bin, crane, ash hopper, ash tunnel, charging hoppers, charging grates, forced draft fans, or, in fact, any mechanical devices are justified. However, auxiliary fuel apparatus is a necessity when wet garbage is being burned.

If, when the design does not warrant auxiliary fuel apparatus and additional fuel is required on rare occasions when material is water-soaked, then a small amount of soft coal or wood will be the most economical solution. A constant source of auxiliary heat involves fixed charges and maintenance.

A plant for areas of from five to twenty tons, might have two furnaces, but will not need a bin and crane, nor an ash tunnel and ash hopper. In the lower brackets, it is highly probable that only one furnace unit should be built.

A plant for twenty to fifty tons may be equipped with either a bin and crane, or a ramp leading directly to the charging floor. A two or three shift operation should be considered in this case. Also, the situation may call for two conveniently located units as the most economical.

Furnace Chamber

The furnace chamber of an incinerator should be designed to promote the rapid desiccation of the raw refuse leaving only the volatile gases to be burned in the mixing chamber and ashes to fall into the ash pit. The grate area in the furnace increases at a decreasing rate with the size of the plant. The relationships used take into account usual air pollution control requirements.

The primary furnace chambers are normally lined with a minimum thickness of nine inches of refractory material and four inches of insulation material. The refractories should be of the best quality available, preferably super duty fire brick. They should be constructed in accordance with acceptable refractory construction methods as specified by the refractory manufacturers, well-maintained, and whenever possible, operated at temperatures above 1000°F. The temperature range, up to 1000°F, is the critical range where maximum expansion and contraction of the expensive refractory material and resulting deterioration takes place. Refractories should be designed to operate at temperatures up to 2400°F, with the optimum burning temperatures maintained in the neighborhood of 1300° to 2000°F within the primary chambers.

It is advisable to install indicating pyrometers in the furnace and secondary combustion chamber for observing temperatures. This results in a minimum of thermal stresses and maximum life, as well as good operating conditions for the primary chamber. The refractories and grates should be installed with future replacement as a consideration.

Flame Port

This is the passageway connecting the furnace and the mixing chamber. A gas velocity of from 45 to 65 feet per second should be maintained at this point.

Mixing Chamber

The mixing chamber is where the volatile gases mix and burn, thus relieving odor and toxicity. Velocities in this chamber should range between 20 and 30 feet per second. The refractories and insulation should be of the same quality and thickness as in the furnace chamber.

Curtain Wall

This is the passageway connecting the mixing chamber and the secondary combustion chamber. Velocities between 15 and 25 feet per second are desirable here.

Secondary Combustion Chamber

In this chamber, also called the expansion chamber, combustion gases expand, burn out, and fly ash settles. Velocities in this chamber should be about 5 to 6 feet per second, and again, high quality refractories should be used. The volume of this chamber should be based on 10 to 12 cubic feet per ton of rated daily (24 hour) capacity, or the velocity, whichever is greater. Fly ash is cleaned from this chamber and the mixing chamber when the plant is shut down for routine cleaning.

In large plants, the use of baffles in the expansion chamber improves the mixing of the gases and increases impingement of the fly ash. Many small plants combine the mixing chamber and the secondary combustion chamber into one.

Flues

A flue is the connection between the secondary combustion chamber and the chimney. In large plants, when more than one furnace is employed, the flues are interconnected so that one chimney can serve two or more units. Since some fly ash is deposited in the flues, they should be readily accessible for cleaning. Curb walls, which function like snow fences, are sometimes used in flues to further stop fly ash from reaching the chimney. The cross-sectional area of a flue usually ranges from one-quarter to one square foot per ton of daily rated capacity. Walls should be so constructed as to be well insulated and both heat and gas tight.

Dampers

Dampers are used to control furnace draft and to cut off draft when the incinerator is closed down. A convenient location for the damper is in the flue just before the chimney. Dampers, being located in a relatively hot part of the incinerator, should be made of steel. (Supported refractory is used on many larger units). Dampers are counterweighted for easy operation.

Chimneys

Chimney dimensions are important to obtain proper draft for the incinerator and disperse the gases over a wide area in a nuisance-free manner. They should be tall enough to clear the highest obstacle. Steel chimneys are recommended for small incinerators.

ENGINEERING DESIGN

This section contains sample calculations used in designing an incinerator. The design factors for either a small or large plant are essentially the same, as the burning process in both cases is the same. Small units may not justify expensive refuse handling equipment, but the engineering design and operating criteria are similar.

Quantity of Refuse

The amount of refuse to be disposed of is a basic consideration in the design and operation of a city's refuse disposal facilities. The amount of municipally collected refuse is the total amount produced less the amount disposed of by on-site methods and nonmunicipal methods.

Until recently there has been a lack of information on refuse generation for large areas within the United States. Most of the information has dealt with selected metropolitan areas that are not necessarily representative of other cities or other areas on a nationwide scale.

Recently, the first national survey of solid wastes was made by representatives of the Solid Wastes Program of the Public Health Service, state agencies, and consultants. The survey was based on a large sample consisting of 92.5 million people (46 percent of the population of the United States) from 33 states.

Weight – Results of the national survey show that approximately 5.32 pounds of solid wastes per person is collected each day. Table 6 summarizes the amount of solid wastes collected in urban and rural areas by category of origin.

Because the urban population is much larger than the rural population, the national averages more nearly approximate the urban average than the rural average.

TABLE 6: AVERAGE SOLID WASTE COLLECTED*

Solid Wastes	Urban	Rural	National
Household	1.26	0.72	1.14
Commercial	0.46	0.11	0.38
Combined	2.63	2.60	2.63
Industrial	0.65	0.37	0.59
Demolition, construction	0.23	0.02	0.18
Street and alley	0.11	0.03	0.09
Miscellaneous	0.38	0.08	0.31
Totals	5.72	3.93	5.32

*(lb/person-day)

Source: PB 200,514 June 1971

Some communities do not collect household and commercial wastes separately, in which case a combined (household and commercial) average is reported.

Volume – The quantity of refuse produced is also expressed as volume. An approximate figure often used is 108 cubic feet of uncompacted refuse per capita per year, which is equivalent to 30 cubic feet of compacted refuse.

Seasonal Influences – Seasonal variations in refuse production can be of great importance in the design and operation of municipal incinerators. Peak loads may be the result of spring cleanup campaigns, autumn leaf collection, or the tourist season in a resort town.

A plot of seasonal variations of the refuse incinerated in two American cities, shown in Figure 5.3 shows two different regimes. Hartford, Connecticut, has an annual cleanup campaign that accounts for the peak in April and May.

In the autumn, great quantities of leaves are collected and incinerated. The minimum during July and August is perhaps a reflection of summer vacations.

The pattern for Cincinnati, where there isn't an organized spring cleanup campaign, indicates that a more gradual increase in refuse occurs with the onset of spring and summer activities.

FIGURE 5.3: SEASONAL DIFFERENCES IN AMOUNTS OF REFUSE INCINERATED*

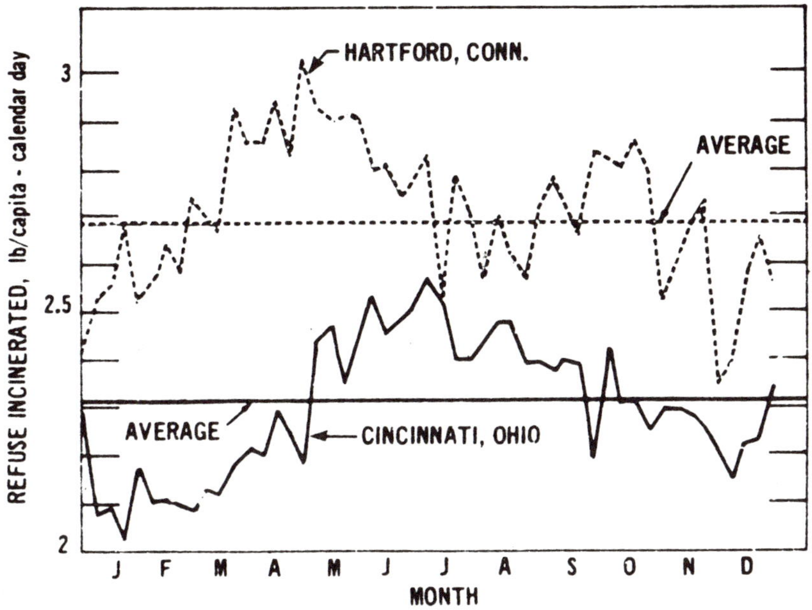

*Hartford, Connecticut, and Cincinnati, Ohio, 1957

Source: PB 200,514 June 1971

Projections — A projection of refuse generation and collections entails many considerations such as changes in packing technology, changes in collection procedures, changes in disposal costs, changes in per capita expenditure for consumption of goods, and perhaps most important, the projection of population.

One of the most recent projections on a nationwide scale, which assumes that per capita waste production increases at a rate similar to the per capita expenditure for consumption of durable and nondurable goods, shows that the amount of material to be collected through municipal and private agencies will rise to 8 pounds per capita per day by the year 1980 based on a 1968 per capita production of 5.32 pounds.

A projection for Kenosha, Wisconsin, (population estimated at 75,000) shows that refuse production will increase from 2.76 pounds per capita in 1965 to 4 pounds by 1976. The per capita projection coupled with the population projection would provide a projection of daily refuse disposal requirements.

Composition of Refuse

The composition of refuse to be incinerated is a major consideration in the design of modern municipal incinerators. The composition of the refuse determines such important quantities as the calorific value, the amount of air required for combustion, the amount of heat released, the characteristics of the exhaust gases produced, and the amount of residue.

Chemical Composition – Refuse can be considered as a combination of moisture, dry combustible material, and noncombustible material. The moisture content of refuse can be either free (visible) or bound (nonvisible). Moisture content fluctuates with the weather, particularly with rain and humidity.

A proximate analysis of 20 of the most common combustible components of municipal refuse (see Table 7) shows that vegetable and citrus wastes are highest in moisture content and paper is quite low in moisture content. All samples showed a high loss of combustible carbon. The fixed carbon is that portion of the refuse that has to be burned out on the incinerator grate.

TABLE 7: PROXIMATE ANALYSIS OF MUNICIPAL REFUSE AS DISCARDED BY HOUSEHOLDERS*

Refuse component	Moisture	Volatile matter	Fixed carbon	Ash	Btu/lb As discarded	Btu/lb Dry basis
Newspaper	5.97	81.12	11.48	1.43	7,974	8,480
Brown paper	5.83	83.92	9.24	1.01	7,256	7.706
Trade magazine	4.11	66.39	7.03	22.47	5,254	5,480
Corrugated paper boxes	5.20	77.47	12.27	5.06	7,043	7,429
Plastic coated paper	4.71	84.20	8.45	2.64	7,341	7,703
Waxed milk cartons	3.45	90.92	4.46	1.17	11,327	11,732
Paper food cartons	6.11	75.59	11.80	6.50	7,258	7,730
Junk mail	4.56	73.32	9.03	13.09	6,088	6,378
Vegatable food wastes	78.29	17.10	3.55	1.06	1,795	8,270
Citrus rinds and seeds	78.70	16.55	4.01	0.74	1,707	8,015
Meat scraps, cooked	38.74	56.34	1.81	3.11	7,623	12,443
Fried fats	0.00	97.64	2.36	0.00	16,466	16,466
Leather shoe	7.46	57.12	14.26	21.16	7,243	7,826
Heel and sole composition	1.15	67.03	2.08	29.74	10,899	11,026
Vacuum cleaner catch	5.47	55.68	8.51	30.34	6,386	6,756
Evergreen shrub cuttings	69.00	25.18	5.01	0.81	2,708	8,735
Balsam spruce	74.35	20.70	4.13	0.82	2,447	9,541
Flower garden plants	53.94	35.64	8.08	2.34	3,697	8,027
Lawn grass	75.24	18.64	4.50	1.62	2,058	8,312
Ripe tree leaves	9.97	66.92	19.29	3.82	7,984	8,869

*Percent by weight

Source: PB 200,514 June 1971

The ultimate analysis for the same samples on a dry basis is given in Table 8.

TABLE 8: ULTIMATE ANALYSIS OF MUNICIPAL REFUSE, DRY BASIS*

Refuse component	Carbon	Hydrogen	Oxygen	Nitrogen	Sulfur	Ash
Newspaper	49.14	6.10	43.03	0.05	0.16	1.52
Brown paper	44.90	6.08	47.84	0.00	0.11	1.07
Trade magazine	32.91	4.95	38.55	0.07	0.09	23.43
Corrugated paper boxes	43.73	5.70	44.93	0.09	0.21	5.34
Plastic coated paper	45.30	6.17	45.50	0.18	0.08	2.77
Waxed milk cartons	59.18	9.25	30.13	0.12	0.10	1.22
Paper food cartons	44.74	6.10	41.92	0.15	0.16	6.93
Junk mail	37.87	5.41	42.74	0.17	0.09	13.72
Vegetable food wastes	49.06	6.62	37.55	1.68	0.20	4.89
Citrus rinds and seeds	47.96	5.68	41.67	1.11	0.12	3.46
Meat scraps, cooked	59.59	9.47	24.65	1.02	0.19	5.08
Fried fats	73.14	11.54	14.82	0.43	0.07	0.00
Leather shoe	42.01	5.32	22.83	5.98	1.00	22.86
Heel and sole composition	53.22	7.09	7.76	0.50	1.34	30.09
Vacuum cleaner catch	35.69	4.73	20.08	6.26	1.15	32.09
Evergreen trimmings	48.51	6.54	40.44	1.71	0.19	2.61
Balsam spruce	53.30	6.66	35.17	1.49	0.20	3.18
Flower garden plants	46.65	6.61	40.18	1.21	0.26	5.09
Lawn grass, green	46.18	5.96	36.43	4.46	0.42	6.55
Ripe tree leaves	52.15	6.11	30.34	6.99	0.16	4.25

*Percent by weight

Source: PB 200,514 June 1971

Carbon is plentiful and is the principal fuel element. Sufficient hydrogen is present in most cases to burn all of the oxygen to water. Nitrogen is present in rather insignificant quantities except in leather, vacuum cleaner catch, lawn grass, and ripe tree leaves. Sulfur is present in most refuse in rather small quantities, particularly when compared to the sulfur content of coal and fuel oils.

Physical Composition – Up to this point we have been concerned solely with the chemical analyses of separate refuse components. These analyses can be used to simulate the total chemical composition for municipalities where accurate estimates of the components can be made. In those areas where component measurements are not feasible or practical, analyses from other municipalities can be useful in making a first estimate.

One such analysis was made for Oceanside, Long Island, New York. Refuse composition and moisture content for three tests are shown in Table 9. Large variations are apparent in both the components and moisture content. Results from these tests and others show that refuse composition and moisture content (as previously stated) both vary from day to day, season to season, on holidays, and with geographical area. The consuming patterns and uniformity of the standard of living throughout the United States are the main factors in creating uniformity in refuse.

A brief review of Table 9 shows that paper products are the main refuse components, generally comprising from 30 to 55 percent, of the total refuse at the Oceanside Refuse Disposal Plant. Grass and dirt obviously showed the greatest variance, ranging from 33 percent in June to essentially zero percent in February. Moisture content varies over a wide range of values depending upon the category of the refuse. The high moisture content of the paper products was explained as being a result of absorption from other refuse components and weather elements.

TABLE 9: REFUSE COMPOSITION AND MOISTURE CONTENT OF EACH COMPONENT

Component	Test 1 June 1, 1966 Weight, percent	 Moisture, percent	Test 2 June 23, 1966 Weight, percent	 Moisture, percent	Test 3 February 21, 1967 Weight, percent	 Moisture, percent
Cardboard	1.59	23.78	6.75	13.22	5.78	16.10
Newspaper	8.88	37.77	11.27	19.20	21.35	18.00
Miscellaneous paper	22.25	36.98	21.78	24.68	26.20	21.00
Plastic film	1.76	18.80	1.77	20.47	1.20	2.85
Other plastics	0.69	20.50	1.67	29.60	2.34	4.38
Garbage	9.58	65.25	10.21	73.45	16.70	59.80
Grass and dirt	33.33	62.20	19.00	44.80	0.26	21.08
Textiles	3.00	31.40	3.33	22.40	2.24	26.05
Wood	1.22	24.98	6.58	8.70	1.46	13.20
Minerals	9.74	6.00	9.49	1.99	11.87	1.64
Metal	7.96	10.83	8.15	2.76	10.60	4.46
Totals	100.00		100.00		100.00	

Source: PB 200,514 June 1971

Refuse Heating Value

The calorific value is considered one of the most important factors of refuse composition for the incineration process. The overall calorific value of refuse is affected by the amount of moisture and the percentages of combustible and noncombustible elements in the refuse.

The heat value of refuse varies widely from country to country. Japanese refuse contains large amounts of moisture and therefore, has a much lower calorific value than refuse in Europe and the United States. The high moisture content in Japanese refuse is due mainly to the presence of more garbage and less paper. The current estimate of the average higher heating value (HHV) of the United

States municipal refuse is 5,000 Btu per pound. Average calorific values may be difficult to determine because values of 2,000 to 10,000 Btu per pound can exist.

The heating value of refuse is normally determined by determining the components and their quantities in a representative sample of the refuse. By applying known heating values to the components, a reasonable calorific value of the refuse can be computed. Such an analysis, based on refuse composition from a number of incinerators during the 1950 to 1962 period, gives a calorific value of 4,917 Btu per pound as-fired. Table 10 gives the composite analysis from which the calorific value (Btu/lb: 4,917 as-fired; 6,203 dry basis; 9,048 dry ash-free basis) was determined.

TABLE 10: AVERAGE MUNICIPAL REFUSE, AS-RECEIVED BASIS

	Percent	Theoretical Combustion Air, lb/lb refuse
Moisture	20.73	
Carbon	28.00	x 11.53 = 3.2284
Hydrogen	3.50 (0.71)*	x 34.34 = 0.2438
Oxygen	22.35	
Nitrogen	0.33	
Sulfur	0.16	x 4.29 = 0.0069
Noncombustibles**	24.93	
Total	100.00	3.4791

*The net hydrogen available for combustion (0.71%) equals the total hydrogen (3.50%) less one-eighth of the oxygen (22.35%/8).

**Ash, glass, ceramics, metals.

Source: PB 200,514 June 1971

In a study of refuse at the Oceanside, New York incinerator, the average calorific values (HHV) of the various components of refuse was computed by using a bomb calorimeter. The results are presented in Table 11.

TABLE 11: HIGHER HEATING VALUES (Btu/lb)

Component	As-Received Basis	Dry Basis	Moisture-and Ash-Free Basis
Cardboard	6,389	7,841	8,131
Newspaper	5,927	8,266	8,518
Miscellaneous paper	5,390	7,793	8,439
Plastic film	11,128	13,846	14,849
Other plastics	6,778	9,049	11,332
Garbage	2,226	7,246	9,287
Grass and dirt	2,970	6,284	9,002
Textiles	5,876	8,036	8,299
Wood	6,850	8,236	8,482
Minerals*	79	84	9,438
Metallic*	683	742	8,439

*Btu in labels, coatings, and remains of contents

Source: PB 200,514 June 1971

Because the heat produced by the oxidation of metals is highly variable among incinerators, it was not included.

Bulk Density

Density is perhaps most important to sanitary landfill oeprations, and will, therefore, not be discussed in detail. (One application of density important to municipal incineration is in determining average grapple loads, a figure that may be used in crane design and the estimation of furnace loading.) Density of refuse as-collected has been decreasing in recent years because of the change in refuse composition. However, this trend may be somewhat affected by some collection vehicles that have been designed to carry larger loads by the use of a compacting device. Density of refuse in an incinerator pit is obviously greater in the lower half than in the upper half because of compaction by the refuse itself.

Studies at the Oceanside Refuse Disposal Plant show that 350 pounds per cubic yard may be a useful figure for pit design. Table 12 gives the average pit densities of refuse for the Oceanside Plant. The lower density and moisture content in March reflect drier weather conditions, absence of dense, high-moisture content yard wastes, and the drying effect of refuse during the domestic heating season.

TABLE 12: PIT DENSITIES OF REFUSE AT OCEANSIDE DISPOSAL PLANT

	- - - - - - - - -Average Density in Pit, lb/yd^3- - - - - - - - -		
	Moisture Content, %	Before Settling	After Settling
March 18, 19	26	349	375
June 13, 14	42	480	523

Source: PB 200,514 June 1971

Sampling Methods

There are apparently no standard procedures used in the sampling of refuse. Most methods are designed to help a city determine what type of disposal method to use, and to determine how the particular disposal system should be designed and operated. If a city has decided to build an incinerator, some characteristics of interest will be the amount, calorific value, density, physical and chemical composition of the refuse. Representative samples are difficult to find because of the heterogeneity of the chemical and physical composition of the refuse. Heterogeneity is introduced by day-to-day changes in composition, weather, and seasonal changes. Preparation of refuse as-received at the incinerator may help provide more representative samples. The Swiss Federal Institute of Technology plans to obtain more representative samples of refuse by the use of a portable hammermill that will grind, homogenize, and mix unsorted refuse.

Incinerator Air Pollution Abatement Equipment

The material for this chapter is based on PB 200,514 June 1971. Incineration of refuse always produces at least two waste products; residue and combustion gases. Except for low "burnout" of the refuse, the residue is usually not a significant disposal problem because it is low in volume, sterile, and its offensive odors have been removed.

The combustion gases, however, can be a significant problem because of their contribution to air pollution. The primary air pollution concern is with particulate emissions rather than gases and odors. At present, air pollution control devices are basically designed for the removal of particulate matter, with some incidental removal of pollutant gases by certain types of control processes. A third waste product, effluent water, can be a problem for municipal incinerators that utilize wet gas-cleaning devices. In areas where water pollution is a major consideration, it may be well to emphasize dry gas-cleaning systems rather than wet systems.

SETTLING CHAMBER (EXPANSION CHAMBER)

A settling chamber, one of the early and simple methods for flyash control, is located immediately beyond the combustion chambers. Large particles of flyash settle out if the gas expansion chamber is large enough in size to substantially lower the gas velocity.

For example, a 30-micron particle settles at the rate of 10 feet per minute, and a 1-micron particle settles at ¼ inch per minute. It is apparent from these figures that, from a practical viewpoint, settling chambers are effective only for the extremely large flyash particles. The chambers are constructed of either refractory brick or steel, and are designed and fitted with devices to keep

FIGURE 6.1: NORTH HEMPSTEAD INCINERATOR

A - DUAL CRANES
B - HOPPER WITH LARGE OPENING
C - FEED CHUTE
D - SAFETY JACKET
E - HIGH-TEMPERATURE REFRACTORY
F - GRATE
G - FURNACE
H - RESIDUE HOPPER
I - SECONDARY COMBUSTION CHAMBER AND DOWNPASS FLUE
J - FINAL BURNING AND SETTLING CHAMBER
K - NOZZLE-CLEARED WET BOTTOM
L - HIGH-PRESSURE OPPOSED SPRAY CURTAIN
M - FLYASH SLUICEWAYS
N - "FAIL-SAFE" DAMPER ARRANGEMENT
O - SEQUENTIAL CYCLONE COLLECTORS
P - FLYASH HOPPER
Q - INDUCED-DRAFT FAN
R - BYPASS FLUE
S - PROVISION FOR ADDED FILTERS OR PRECIPITATORS
T - CHIMNEY

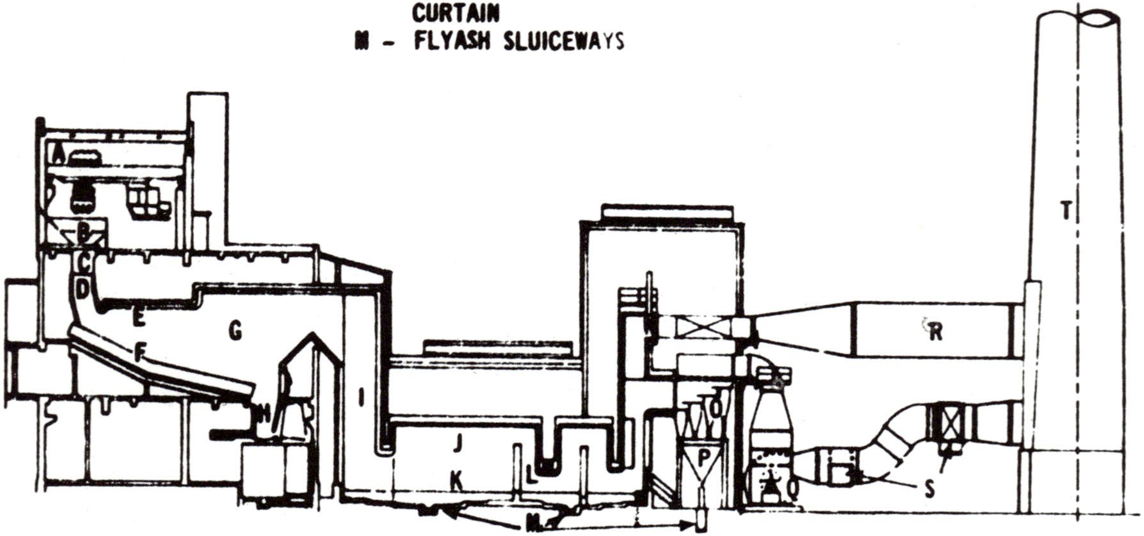

Special features include an unusual amount of air pollution control equipment, provision for additional equipment if needed, and an emergency bypass flue.

internal turbulence to a minimum to keep flyash from becoming reentrained in the gas stream. Reentrainment can also be reduced by using a wet bottom chamber. Deflecting dampers are installed in some wet bottom chambers to force the flyash laden air against the water surface. Gravity settling is effective only for particle sizes of 200 microns or more, and settling chamber efficiency usually averages only 15 to 25%.

Such chambers are therefore desirable only for the removal of large particles prior to further cleaning by more sophisticated devices. Such a scheme is used in the North Hempstead incinerator, which uses two other gas-cleaning devices in addition to a wet bottom settling chamber (see Figure 6.1). In anticipation of more stringent codes, provisional space was provided for more efficient cleaning equipment.

BAFFLED COLLECTORS

In some incinerators, baffled collectors are installed separately from settling chambers. They are usually made of brick or metal and can be either wet or dry. There are many collector designs and collectors can be placed in several locations within the postcombustion chamber area of the incinerator. Particles are removed by direct impingement, velocity reduction, or centrifugal action. Removal efficiencies are quite low and only larger flyash particles, mostly 50 microns or larger can be removed.

SCRUBBERS

Scrubbers clean the combustion gases by carrying wetted flyash to the bottom of the scrubber. To be incorporated into the water, flyash particles must impact on a water droplet. The impaction efficiency is primarily a function of the relative velocity between the flyash particle and the water droplet, the size and density of the flyash, the number of water droplets, and the fineness of the water spray. Most of these factors are a function of the pressure drop in the scrubber and the energy input to the scrubber system.

In some incinerators, fresh water is used continuously for the scrubbing process, which necessitates disposal of the slurry leaving the scrubber. Because the amount of water required by scrubbers is high, however, and economy is of interest, the scrubber water can be recirculated after removal of the wetted flyash. Corrosion from the acidity of the scrubber water, caused by the absorption of acid-forming combustion gases, can be a serious design and maintenance problem. Some installations use the slurry for quenching the hot residue as it falls from the burning grate. After the quenching, the slurry may or may not be recirculated.

Scrubbers are usually made of stainless and carbon steels. Maintenance usually consists of repair and replacement of spray nozzles or flow valves. Efficiency is related to the pressure drop in that higher efficiencies require higher pressure drops. In the venturi scrubber, pressure drops of from 20 to 40 inches of water can be required.

White stack plumes are common, particularly in cold weather when efficiently scrubbed gases are laden with large amounts of moisture added during the scrubbing process. Indicative of moisture, rather than pollutants, the plume has the appearance of being a pollutant, and codes with given opacity requirements can require elimination of such a plume. The most obvious method for elimination of the steam plume is to use a dry gas-cleaning method.

Many methods have been suggested for suppression of the steam plume. Some of them are: electrostatic precipitation of the water droplets, mechanical separation of the water droplets, absorption or adsorption of water vapor, mixing of the moist gases with relatively dry heated air, condensation of the moisture by direct contact with water on cold surfaces, and reheat of scrubber exhaust gases. Studies show that costs of these steam plume suppression methods, when the ambient air temperature is above 20°F can be as much as the cost of a wet scrubber system.

Spray "Walls"

Perhaps the simplest scrubber design (some may question its classification as a scrubber) consists of a water spray wall in which the spray is arranged to permit maximum contact between the water and the dirty gas. The sprays can be placed in several locations, such as the settling chamber, the baffle collector, the breaching ducts, or a chamber specifically designed for scrubbing.

Venturi Scrubber

Both flyash and gaseous pollutants are removed in a venturi scrubber in which water is supplied peripherally at the top of the venturi. Gases passing through the venturi tube are accelerated at the throat to a velocity that fragments the water into a mass of fine droplets. Impaction efficiency is high because of high relative velocities, small water droplet size, and large number of droplets in the throat of the venturi tube. Downstream from the throat, the cleaned gases decelerate and the water droplets agglomerate to a size easily separated from the gas stream.

Venturi scrubbers have a high collection efficiency, usually 90% or greater, and can process untreated gases directly from the combustion chamber. Recirculation of scrubber water permits less consumption of water and ensures a minimum, but concentrated production of slurry.

Cyclonic Spray Scrubber

The configuration of one style of cyclonic spray scrubber is illustrated in Figure 6.2. Gases enter the lower portion of the scrubber peripherally and make a helical motion through a water spray until they exit the top of the scrubber. Water is supplied to a spray manifold, which is located in the center of the scrubber. Slurry is drained from the bottom of the scrubber. Impaction efficiencies here depend on the velocity of the gases and the atomization of the water by the spray manifold. Efficiencies of cyclonic scrubbers range from 85 to 94%.

FIGURE 6.2: CYCLONIC SPRAY SCRUBBER

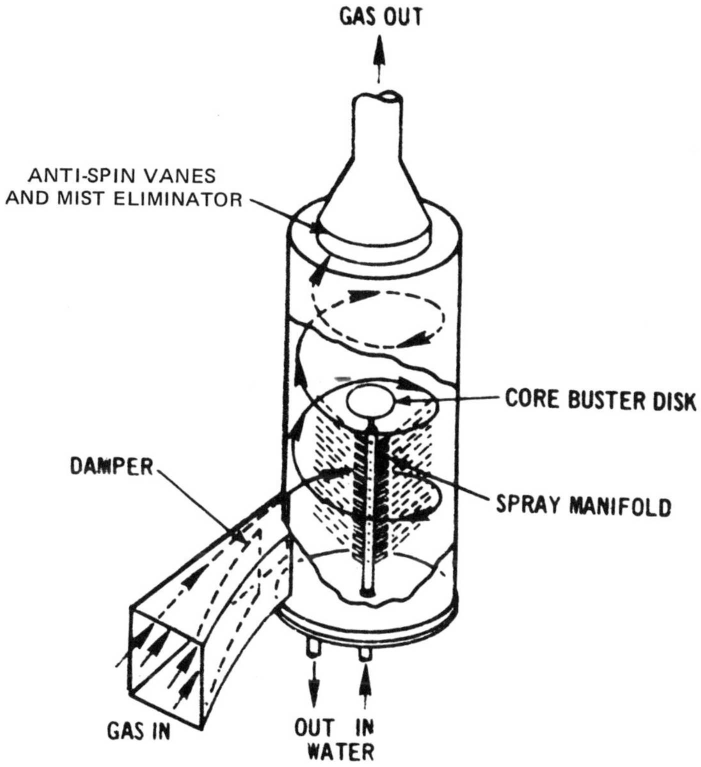

Packed Scrubber

Packed scrubbers clean gases by the same process as other scrubbers but achieve impaction of the flyash with water in a unique way. Gases enter the cover portion of the scrubber (see Figure 6.3) and pass through a series of packed beds that are wetted from the top by water sprays. Flyash is carried to the bottom of the scrubber where it is removed.

FIGURE 6.3: PACKED SCRUBBER

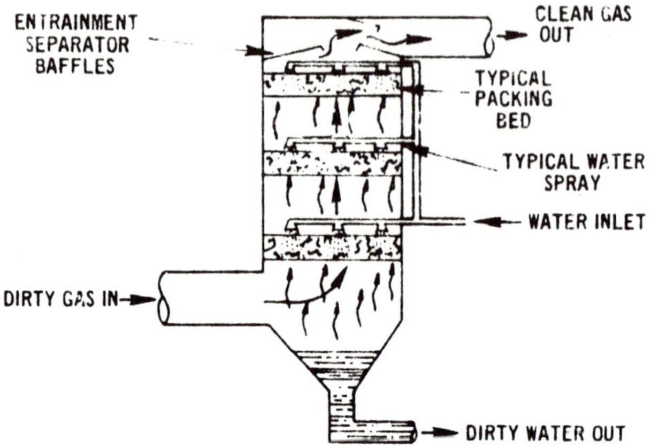

Flooded-Plate Scrubber

Another scrubber type is the flooded-plate scrubber shown in Figure 6.4. Gases enter the bottom of the scrubber and pass through a series of water-flooded plates containing a myriad of water-covered holes. Clean gases exit at the top of the scrubber and the slurry is drained from the bottom. Collection efficiency ranges from 90 to 95% and water requirements range from 3 to 5 gallons for 1,000 cubic feet of gas treated.

FIGURE 6.4: FLOODED-PLATE SCRUBBER

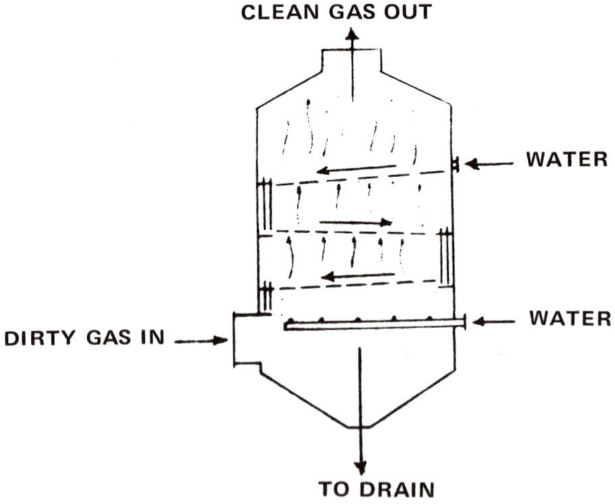

CYCLONE COLLECTORS

Cyclones are able to remove particulate matter from the exhaust gases without the use of water by means of centrifugal separation of the particles and gases. There are two basic types of cyclone collectors, the multicyclone and the involute cyclone (Figure 6.5). Gases must be cooled to within the range of 400° to 700°F to permit standard construction of cyclone collectors and induced-draft fans.

FIGURE 6.5: CYCLONE DUST COLLECTOR, INVOLUTE

Multicyclone Collector

Polluted gases enter the collector through a spinning vane, which sets up an intense vortex. Particles are centrifugally thrown against the walls of the collector and fall to the bottom where they are removed. The cleaned gases exit vertically through the outlet tube. The tubes of a multicyclone collector are usually from 9 to 10 inches in diameter and are mounted in two common tube sheets. One sheet is for incoming dirty gases and the second tube sheet is for the exit of the cleaned gases.

Multicyclones are more efficient for larger particles than they are for smaller particles. Efficiency drops off rapidly for particles smaller than 20 microns. For 10-micron particles, only 35% (by weight) can be collected. For a pressure drop of 3.5 inches of water, a multicyclone collector can obtain an efficiency of about 80%. Plugging of the cyclone, which can be a serious problem in this type of collector, can lower efficiency significantly.

Involute Cyclone

Involute cyclones, which are much larger than multicyclones, are usually 2 to 5 feet in diameter. They operate on the same principle as the multicyclone, but are not subject to the plugging. Erosion of the lower cone can occur and is usually corrected by ceramic lining or water flushing to remove collected flyash. One particular model of the involute cyclone solves this problem by introducing water peripherally at the top of the cyclone wall. An additional advantage of this design is that it permits the water to carry the flyash down the walls, eliminating reentrainment of the ash.

In addition, the temperature of the gas may be reduced by as much as 200°F by passing the gas through a wet cyclone. The moisture added to the gas is usually not sufficient to create a steam plume.

FABRIC FILTER COLLECTORS

Fabric filter collection of flyash from incinerators has potential of being an effective and appropriate method of flyash control, but at present the method is still in its preliminary stages of development.

FIGURE 6.6: FABRIC FILTER DUST COLLECTOR

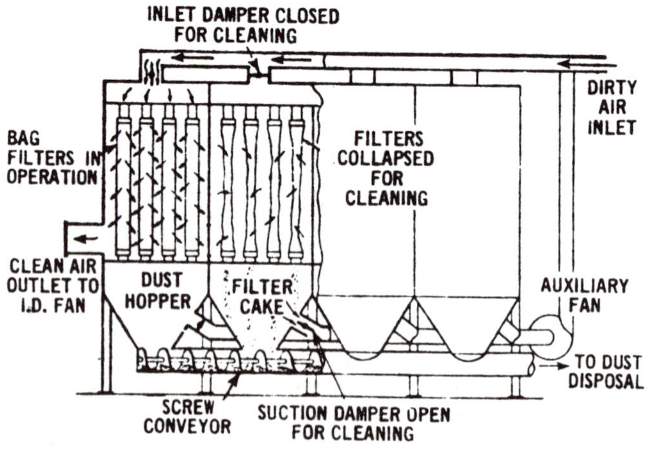

Slow development can be attributed to the high-temperature gases that must be filtered and the characteristics of the flyash. The efficiency of fabric filtration is high. Fabric filters cannot be overloaded during periods of excessive dust loadings as is common with other types of control equipment. Tests of fabric filters installed at an incinerator in Pasadena, California, show that they have an efficiency of 99.77%. The filters are usually arranged as tubular bags so that they can be cleaned by shaking, bag collapse, reverse jet blowing, and reverse flow backwash. The bags are connected to a dust hopper into which the caked dust falls for removal (see Figure 6.6).

Pressure losses for bag filters range from 3 to 7 inches of water. Glass fabrics used can withstand temperatures up to 500°F. A disadvantage of fabric filters is that they require more room than any other one air pollution control device. The initial cost of such filters and their maintenance is high. Extensive gas conditioning and control is required for proper performance of the filter fabric.

ELECTROSTATIC PRECIPITATOR

What is perhaps the first electrostatic precipitator installed in an incinerator was installed in the late 1920s in Zurich, Switzerland. Since that time electrostatic precipitators have become widely used throughout most of Europe. All municipal incinerators located in large cities in Japan use electrostatic precipitators for gas cleaning. England has used electrostatic precipitators in only a few of its incinerators to date, but the Deephams' Refuse Disposal Works in London, with a planned capacity of 1,667 tons per day, will be equipped with electrostatic precipitators.

The first two incinerators in the United States to be equipped with electrostatic precipitators are located in New York City. As a pilot project, two precipitators are supplied by two different manufacturers and placed in two different existing incinerators. Only one furnace at each of the two New York incinerators will be equipped with the new cleaning device. The city of Montreal, Canada is following the lead of the Europeans in the design of a 1,200-ton-per-day capacity incinerator that will be equipped for steam production and will use an electrostatic precipitator for each of its four furnaces.

Operating Principles

The basic process by which electrostatic precipitators separate dust or moisture from a gas stream is relatively simple. An electrostatic precipitator consists of discharge wires of relatively small diameter and collecting surfaces, such as plates or tubes, between which gases pass carrying entrained particles. The discharge wires are the pole of negative polarity while the collecting surfaces are positive and at ground potential.

A unidirectional, high-potential field is set up between them. At and above a critical voltage, a corona discharge takes place near the surface of the negative wire. The corona is a visible manifestation of the ionization of the gas between the poles resulting in the formation of positive and negative gas ions in the region near the negative wires. These ions are attracted to the pole of opposite polarity.

In moving toward the opposite pole the ions attach themselves to the dust particles entrained in the gas, charging the particle positive or negative as the case may be. The particles themselves are then attracted to the pole of opposite polarity on which they are deposited. Since the ions are formed in the immediate vicinity of the negative wire the negative ions have a much longer distance to travel; hence, more entrained particles are charged negative than positive, resulting in a greater collection on the positive collecting surfaces than on the negative wires.

On reaching the collecting surface, the particles give up their charge and adhere to it lightly until dislodged by rapping. From this it can be seen that there are four steps for electrostatic precipitation: (1) charging the particles by means of gaseous ions or electrons; (2) transporting the charged particles through the gas to the collecting surfaces; (3) discharging the charged particles; and (4) removing the precipitated material from the wires and collecting surface.

Combustion Gas Conditioning

Combustion gas must be properly conditioned prior to entering electrostatic precipitators. Design criteria of precipitators limit the temperature of the inlet gases to a range of from 450° to 600°F. Another factor is the efficiency of a precipitator, which is related to the temperature and moisture content of the gases.

TABLE 1: COMPARISON OF CONDITIONING SYSTEMS

System	Water only	Water and air
Incinerator exhaust		
Gas volume, cfm	169,500	169,500
(250-ton furnace)	at 1310°F	at 1310°F
H_2O dewpoint, °F	104	104
Dust load, g/acf	0.241	0.241
Conditioning system		
Spray water, gpm	80	40
Ambient air, cfm	–	45,400 at 68°F
Precipitator inlet		
Gas volume, cfm	130,150 at 560°F	205,200 at 572°F
H_2O dewpoint, °F	150	120
Dust load, g/acf	0.314	0.20
Precipitator size	x	x times 1.77
Precipitator outlet		
Gas volume, cfm	130,150 at 560°F	205,200 at 572°F
H_2O dewpoint, °F	150	120
Dust load, g/acf (Residual dust)	0.018	0.010
Power required – Precipitator fans, pumps	350 kW	640 kW

In installations using waste heat boilers, the temperature reduction is handled by the heat recovery process. Installations not incorporating a waste-heat recovery process can condition the gases in three ways. The first system is a separate evaporation cooling tower that is installed immediately following the last combustion chamber. Cooling of the gases is accomplished entirely by water. The second method is an air-water system in which air and water infiltrate and cool the gases. The third system cools entirely by water that is injected at the end of the furnace rather than in a separate cooling tower. A comparison of the "water only" and "water and air" conditioning systems for a typical 250-ton furnace is given in Table 1.

The table indicates that the water and air conditioning system results in 57% more gas volume to be treated. This factor, in addition to the effect of the dew point on the precipitation process in the water only system, results in a precipitator that is 77% larger for a water and air system than for a water only system. The dust load at the precipitator outlet for the water and air system must be lower to account for the effect of the "diluting air."

Efficiency

Electrostatic precipitators can be designed for nearly any efficiency required, with a pressure drop of only 0.5 to 1 inch of water. Experience with precipitation of flyash from American refuse incineration is limited to only one pilot plant for which test results yielded a collection efficiency (by weight and 50% excess air) of up to 94.4%. The precipitators that will be installed in New York are designed for 95% efficiency. In Europe, because of more stringent codes or the anticipation of more stringent codes, installations with guaranteed efficiencies of over 99% are common.

Table 2 presents some of the basic design elements for 52 precipitator units that are installed in 27 European incinerators. The range of efficiency is 92.0 to 99.5%, but the average is 98.0%.

TABLE 2: DESIGN ELEMENTS OF EUROPEAN ELECTROSTATIC PRECIPITATORS*

Characteristic	Range	Average median
Size of furnaces, tpd	42 to 1,060	270
Raw gas volume, cfs	350 to 7,200	1,450
Dust load in raw gas, lb/1,000 lb	2.7 to 12.3	5.43
Gas entry temperature, °F	285 to 520	490
Overall particulate cleansing efficiency, percent	92.0 to 99.5	98.0

*27 incinerator plants, 52 precipitator units.

Physical Characteristics

Electrostatic precipitators are quite large. Except for fabric filter baghouses, they are perhaps the largest control devices used for municipal incineration. In view of the increased concern with air pollution, when such devices are not made a part of the original design, designers of new plants should consider making allowances for their future installation.

COMPARISON OF AIR POLLUTION CONTROL EQUIPMENT

Innumerable comparisons may be made among air pollution control devices. Perhaps in the end, the single most important element of a control device is its collection efficiency. From previous discussions, perhaps a relative rating of efficiencies can be surmised. Figure 6.7 presents the ranges of collection efficiency for the various classes of control devices. This figure also presents the stack emissions for a given dust loading and collector efficiency.

FIGURE 6.7: COLLECTOR EFFICIENCY VERSUS STACK DUST EMISSIONS

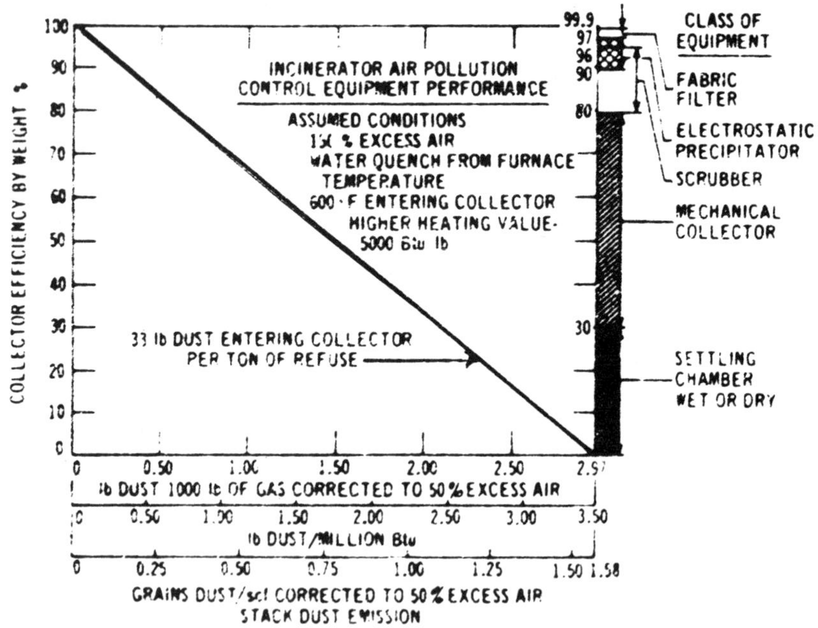

Costs of control equipment are difficult to estimate because of variations among manufacturers, the effect of efficiency and reliability on the cost, and differences in design permitted by local climate. Gas-cleaning costs generally depend on the amount of excess air used, the inlet gas temperature, and the heating value of the refuse. One of the most recent rule-of-thumb estimates of costs of gas-cleaning equipment is presented in Table 3. These figures are based on collectors constructed of mild steel. It has been suggested that because of the difficulties encountered with the collection of flyash from incinerators, air pollution control equipment costs would be found in the higher ranges given below.

TABLE 3: ESTIMATED COSTS OF GAS-CLEANING EQUIPMENT ($/cfm)

Type of collector	Equipment	Erection	Yearly maintenance and repair
Mechanical	0.07 - 0.25	0.03 - 0.12	0.005 - 0.02
Electrostatic precipitator	0.25 - 1.00	0.12 - 0.50	0.01 - 0.025
Fabric filter	0.35 - 1.25	0.25 - 0.50	0.02 - 0.08
Wet scrubber	0.10 - 0.40	0.04 - 0.16	0.02 - 0.05

The economics of decrease in price per volume of gas treated with increase in unit size is reflected in this table. Table 4 compares the relative cost, space, efficiency, water usage, pressure drop, and operating costs for the various types of incinerator air pollution control equipment. It can be seen that in some instances initial cost savings can be lost to high operating costs. High pressure drops mean higher costs because of increased fan loading. Water usage or space requirements can make an otherwise attractive device become quite unattractive.

TABLE 4: COMPARATIVE AIR POLLUTION CONTROL DATA FOR MUNICIPAL INCINERATOR

Collector	Relative capital cost factor (F.O.B.)	Relative space, %	Collection efficiency, %	Water to collector, gpm/1,000 cfm	Pressure drop, in. water column	Relative operating cost factor
Settling chamber	Not applicable	60	0 - 30	2 - 3	0.5 - 1	0.25
Multicyclone	1	20	30 - 80	None	3 - 4	1.0
Tangential inlet Cyclones to 60-in. diameter	1.5	30	30 - 70	None	1 - 2	0.5
Scrubber	3	30	80 - 96	4 - 8	6 - 8	2.5
Electrostatic precipitator	6	100	90 - 97	None	0.5 - 1	0.75
Fabric filter	6	100	97 - 99.9	None	5 7	2.5

All of these estimates and comparisons are first approximations that must be used only in that respect. More meaningful estimates can be quoted by architects, engineers, and equipment manufacturers when a certain type of incinerator is under consideration.

Corrosion in Incinerators

The material in this chapter is based upon PB 213,378, 1972, prepared by the Battelle Memorial Institute for the National Enviromental Research Center.

The use of pollution control devices requires that the incinerator furnace gases be cooled to permit operation of the flue-gas cleaning devices. An economically attractive way in which to do this is by absorbing much of the heat by water contained in wall tubes and in convection-pass tubes used to form the furnace. Alternate procedures are to dilute the hot gases with cool air or with water. However, the latter practices greatly increase the gas volumes to be handled and thus increase the size of the pollution control equipment.

In addition to greatly decreasing the size requirements for pollution control equipment and fans, the waterwall incinerator has several other attractive features. First, there is a gainful use of the heat energy available in the refuse. Second, the furnace throughput can be increased because of the rapid absorption of heat possible. Third, wall slagging problems often encountered in refractory-type construction are absent.

The technology of the waterwall incinerator was developed first in Europe where it is in fairly extensive use. Corrosion problems have been reported, however, in some instances. The first operating water wall unit on this continent was at the Navy Public Works Center in Norfolk, Virginia, beginning in 1969. Others are now operating in Montreal, Quebec; Braintree, Massachusetts; and northwest Chicago. It is anticipated that this type of construction will become used more and more. It is important to anticipate problems which may arise before use begins.

One problem faced in the operation of some European incinerators using water cooled tubes in the walls is the unpredictably severe loss of metal from the furnace tubes. Since usage of large incinerators is increasing in the United States, it is important that the problem be understood so that techniques for its mitigation can be introduced.

As part of a study, probes were placed in a refractory lined municipal incinerator at Miami County, Ohio, and in a waterwall incinerator operated by the Navy Public Works Center at Norfolk, Virginia. Exposure time at temperature ranged from 122 to 1,318 hours. The eleven alloys listed below were used to form the individual specimens which made up the corrosion probes.

TABLE 1: PERFORMANCE OF ALLOYS IN FIRESIDE AREAS*

Alloy	Resistance to Wastage: 300° to 600°F	600° to 1200°F	Moist Deposit
Incoloy 825	Good	Fair	Good
Type 446	Good	Fair	Pits
Type 310	Good	Fair	SCC**
Type 316L	Good	Fair	SCC
Type 304	Good	Fair	SCC
Type 321	Good	Fair	SCC
Inconel 600	Good	Poor	Pits
Inconel 601	Good	Poor	Pits
Type 416	Fair	Fair	Pits
A106-Grade B (Carbon Steel)	Fair	Poor	Fair
A213-Grade T-11 (Carbon Steel)	Fair	Poor	Fair

*Arranged in approximate decreasing order
**Stress-corrosion cracking

Metal wastage rates were shown to be temperature dependent; for the carbon steels they ranged from about 10 mils per month at 325°F to about 35 mils per month at 950°F. These values, however, should not be projected to exposures of many months or years since rates decrease with extended exposure. It is estimated that rates for carbon steel tubes at about 500°F might be expected to range below about 10 mils per year for extended exposure periods.

The thermal gradients between the tube metal and the hot gases were found to be important in addition to the metal temperature.

The tabulation shows the relative performance of eleven materials studied. As an approximation it can be assumed that poor is less than 20 mils per month, fair is 10 to 20 mils per month, and good is more than 10 mils per month. It can be seen that the Incoloy 825 gave encouraging results over the entire temperature range and is much more resistant to stress corrosion cracking than the austenitic stainless steels.

The stainless types, except Type 416, lose less weight than the carbon steels at comparable temperatures, with Types 310 and 446 being superior. The major limitation for these stainless alloys is stress corrosion cracking or pitting when in contact with moist deposits as during downtime. The Inconel 600 and 601 materials were resistant over the lower temperature region but were severely attacked at high temperatures. While aluminum, chromium, and inorganic coatings furnish some initial protection to the tube surfaces they do not appear to be sufficiently durable to be considered for long-time boiler operation.

It was concluded that waterwall incinerators should not be used to generate high temperature superheated steam but should be operated at relatively low metal temperatures, near 500°F, to minimize tube wastage. Steel tubes of compositions such as A106-Grade B or A213-Grade T-11 can be considered at these temperatures. The Incoloy 825-type alloys would be better choices at higher temperatures.

The composition of the deposits removed from the corrosion specimens was carefully determined to provide a clue as to possible causes of the corrosion. Significant amounts of chloride, sulfur, lead, zinc, potassium and calcium, along with other more inert elements, were detected in the deposits from the two incinerators mentioned and also in incinerators at Oceanside, Long Island, and Atlanta, Georgia. Another interesting observation was that the lead content in the deposits at Norfolk and at Atlanta was unusually high.

The proportions of the various chemical compounds found on the tube surfaces depended on the probe specimen temperature, and the distributions were consistent with known volatilities of the compounds.

Composition of the flue gas was determined at the first three sites mentioned. The HCl and SO_2 were present in sufficient concentrations (10 to 300 ppm) at all sites to be of concern and to account for the types of deposits found on the tubes.

CHEMICAL SPECIES RESPONSIBLE FOR CORROSION

Furnace-Gas Compositions

Understanding of the mechanism by which deposits form and corrosion occurs on the metal surfaces of the incinerator, among other things, requires a knowledge of the furnace-gas composition. Sampling of these gases has been done at the Miami County, Ohio, the Oceanside, New York, and the Norfolk, Virginia, sites.

As would be expected from the variable nature of the solid waste burned in municipal incinerators, a wide range of values for the different gases was found. The data are summarized in Table 2. Two gases of primary concern from the corrosion standpoint are SO_2 and HCl. The amounts of these gases found in each of the three incinerators varied greatly. The maximum values are not much over 100 ppm, except for one measurement for HCl of 330 ppm at the Oceanside incinerator and 300 ppm SO_2 at Miami County. In general, the SO_2 and HCl concentrations at Norfolk were less than those at the other sites.

These lower gas concentrations could have some relationship to lower corrosion experienced at Norfolk, although factors such as temperature are probably of even greater importance.

TABLE 2: FURNACE-GAS COMPOSITIONS

	Concentration Ranges		
	Miami County	Oceanside	Norfolk
SO_2, ppm	2-303	0-100	0-10
HCl, ppm	5-115	1-330	14-75
HF, ppm	Traces-0.6	Traces-1.1	Traces-6.0
Nitrogen oxides, ppm	10-138	107*	4-7
Organic acids, ppm	35-136	Not measured	150-340
Oxygen, volume percent	15.0-9.5	17.0-8.0	16.5-15.0
CO_2, volume percent	6.0-10.8	3.5-12	3.9-5.6
CO, volume percent	Traces-0.2	0.005*	0.1-0.2

*Single measurement

It can be noted in Table 2 that the SO_2 concentrations are relatively low compared with those for large coal- and oil-fired boilers, which usually range from 2,500 to 3,000 ppm. On the other hand, such boilers do not contain the HCl found in incinerators. Even at low SO_2 levels, accumulation of sulfates in the incinerator deposits reach the same levels (20 to 30 weight percent) in a few hundred hours as are commonly found in power-station boilers after thousands of hours of service.

Although the HCl levels in the incinerator gases are generally of the same magnitude as the SO_2 levels, the amounts of chlorides found in the deposits are smaller than the amounts of sulfates, presumably because of the conversion of chlorides to sulfates and loss of the chlorine. SO_3 was not believed to be a significant factor. One such measurement at Miami County showed 1 ppm SO_3 and 303 ppm SO_2 in the flue gases at the position where a probe was exposed.

It is concluded that SO_2 and HCl were present in sufficient concentrations in all three incinerators to be of concern from the standpoint of possibilities of corrosive conditions. The amount of HF found at Norfolk (6 ppm) is appreciably greater than the amounts detected at the other two sites. No indications have been found that this gas is of great significance in the tube-metal-wastage process.

While the nitrogen oxides and organic acids contribute to the acidity of the furnace gases, they probably play larger roles in scrubber corrosion, where they are absorbed in wash water, than in fireside corrosion. Laboratory experiments have shown that formic acid vapors have negligible corrosion effects on the steels used.

Relation of Solid Waste to Gas Composition

The composition of furnace gases were observed during six consecutive operating days at the Miami County incinerator in April, 1970. A checklist of typical classes of materials in the solid waste was drawn up, and the presence of a relatively large amount (5 percent or more) of any item was noted. At the same time, the gases in the incinerator were analyzed for SO_2, HCl, HF, nitrogen oxides, organic acids, and particulates. A summary of the data, with duplicate analyses in some cases, is presented in Figure 7.1.

FIGURE 7.1: COMPARISON OF FURNACE-GAS COMPOSITION AND SOLID WASTE AT MIAMI COUNTY INCINERATOR

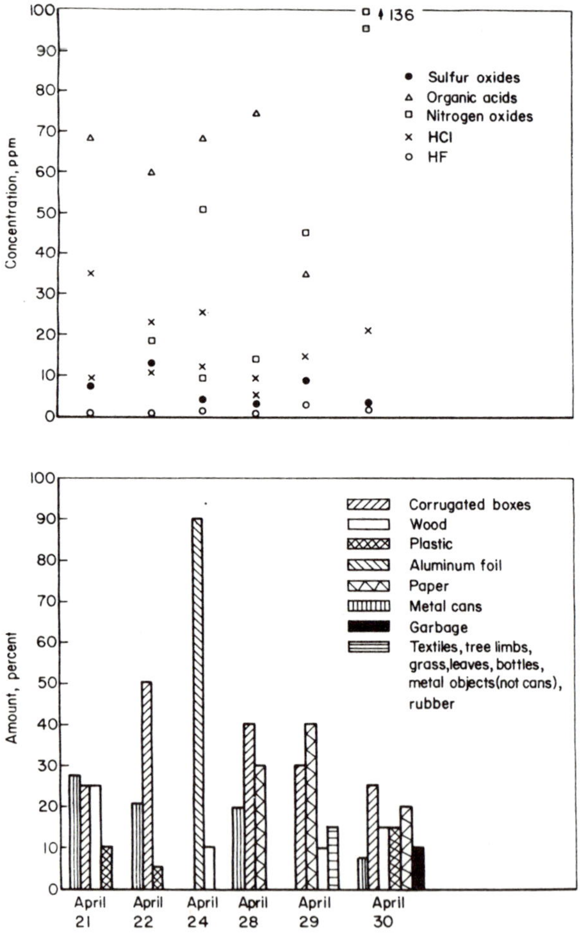

The upper half of the figure shows the concentrations of the gaseous components of the furnace gases while the lower half shows the amounts of various materials in the hopper fed to the incinerator during the sampling period. Any materials present in amounts of 5 percent or more are included in the figure. Lesser amounts of material could not be estimated, so the totals for each day are slightly less than 100 percent.

Although it is difficult to draw any firm conclusions from the nature of the solid waste and the furnace gases, some observations can be made:

(1) Organic acids were the predominant component on four of the days; on these days, corrugated boxes, paper products, or wood were present in large amounts.

(2) The amount of HCl found was significantly below the 100 ppm level which was observed on several other occasions. The presence of 5 to 10 percent of plastics in the waste did not cause any large increase in HCl concentration. The type of plastic being burned was not identifiable, however.

(3) The very small amount of HF detected on these days agrees with that found on other occasions, and probably is typical.

(4) The large amount of nitrogen oxides observed on the last day may be related to the fact that this was the only day on which garbage was present in a notable amount.

In order to obtain a reliable correlation between the solid waste being burned and the nature of the furnace gases, it would be necessary to add known amounts of various materials and determine the changes in the furnace-gas components.

CORROSION FROM INCINERATOR DEPOSITS UNDER HUMID CONDITIONS

All incinerators must be shut down on occasion and high humidity conditions can often cause the deposit on the tube surfaces to become moist. The most damaging forms of attack to be anticipated under these conditions are stress corrosion cracking (SCC) and pitting. Probably the most hazardous of these is stress corrosion cracking.

When an alloy is simultaneously subjected to surface tensile stresses and a variety of specific corrosive environments, it may experience premature failure because of stress corrosion cracking. This process is unique in that failure does not occur when the individual factors mentioned above are acting alone. One aspect of the phenomenon is that the metal at the failed areas does not exhibit ductile tearing, but instead exhibits a brittle fracture. Crack propagation, once initiated, is quite rapid. The propagation is in a plane perpendicular to the direction of the applied stress. The tensile stresses required for cracking may be either applied or residual. The latter may result from welding or from press fits or other assembly procedures. Also, the presence of crevices may favor the initiation of stress corrosion cracks.

All common structural materials such as carbon steel, low alloy steels, stainless steels, aluminum alloys, copper alloys, and nickel alloys are susceptible to SCC to varying degrees in some specific environments. The mechanism of the cracking is not completely understood. From an engineering standpoint, however, it is important to know which combinations of materials and environments will initiate cracking.

Mild steels will crack in certain nitrate, caustic, and carbonate-bicarbonate aqueous solutions and in some gaseous CO-CO_2-H_2O mixtures, anhydrous ammonia, and H_2S-containing environments. Obviously, these substances cannot all be avoided in incinerators. It is considered unlikely, however, that the specific combinations of these substances which promote SCC of mild steel will be encountered in incinerators except on rare occasions. Aqueous nitrates and gaseous CO-CO_2-H_2O mixtures would be the most likely causes of SCC in mild steel incinerators.

While SCC of stainless steels can occur in caustic environments, the most important substance promoting the SCC of stainless steels is chloride ion. Trace quantities of this substance are enough to cause austenitic stainless steels to develop stress corrosion cracks rapidly. While higher temperatures generally favor SCC, Type 304 will crack even at room temperature. Stainless steel alloys will crack when stressed to as low as 2,000 psi in the presence of a few parts per million of chloride ions.

Two groups of materials were evaluated for SCC while in contact with incinerator deposit under humid conditions in the Battelle studies. The first group of alloys included Types 304, 310, and 446 stainless steels and carbon steels A106, Grade B, and A213, Grade T-11. The Group 2 specimens included Type 316L stainless steel and three other alloys more resistant to SCC in chloride environments, namely, Inconel 600, Inconel 601, and Incoloy 825. Bent specimens (C-rings) were stressed to a level halfway between the yield stress and the ultimate tensile strength (at the SCC test temperature). Some U-bend configurations were also used. Some specimens of Type 304 stainless steel were sensitized for 2½ hours at 1200°F to increase their susceptibility to SCC.

In the Group 1 alloys, the Type 304 specimens (both sensitized and annealed) showed cracks after an exposure of one week. Fracture occurred on the sensitized Type 304 piece during the eighth week of exposure. Cracks were first observed in Type 310 steel after 10 weeks. Deep stress corrosion cracks developed in all three materials during 20 week's exposure. There were fewer cracks on the Type 310 specimens but they were very deep. On the other hand, no SCC has been observed with Type 446 steel after 22 weeks of exposure. The specimen, however, was quite rusty and pitted. After 12 weeks, the A106 and T-11 steels were quite rusty and scaled but no stress corrosion cracking was observed. The surface roughening on these steels was noticeable at 15X but was not great. There was no indication of deep pitting as is sometimes seen in corrosive environments. The Group 2 alloys were exposed for 26 weeks. Only Type 316L specimens exhibited cracking and this was at 22 weeks, however some rust appeared in 4 weeks. At 26 weeks, the Inconel 601 specimen showed many fine, shallow pits. Some of these appeared at four weeks as rusty spots. The Inconel 600, at 26 weeks, was mottled in appearance as if it were etched in spots. This alloy also showed a few rusty spots after four weeks.

The Incoloy 825 showed good resistance throughout the 26 week period. It is, therefore, concluded that this alloy is a good choice when resistance to downtime conditions is considered. Since the steels A106, Grade B; and A213, Grade T-11 are corroded in a fairly uniform manner under these conditions, they can also be considered for construction materials. The relatively good performance of the tubes in the boiler at the Navy Public Works Salvage Fuel Boiler provides additional evidence for consideration of carbon steels.

The austenitic type stainless steels such as 304, 310, 316, etc., do not appear to be good choices because of the likelihood of the incidence of stress corrosion cracking. The ferritic type stainless steels as exemplified by 446, have the limitation of possibly being subject to severe pitting under the downtime conditions. These laboratory conditions were more severe than those experienced in the field. Ferritic steels like Type 446 are thought to be immune to SCC in chloride media, but they may be embrittled by hydrogen, particularly when sulfur is present.

Stressed C-ring specimens of Types 304, 310, and 446 steel have also been exposed to pulverized incinerator deposit at 1000°F for 50 and 250 hours in a simulated incinerator atmosphere containing HCl and SO_2. None of the four stainless steel specimens developed any evidence of SCC under these conditions. However, all four specimens were severely corroded on the surfaces that contacted the deposit.

Other stressed C-ring specimens were exposed at 1000°F for 50 and 250 hours to a salt mixture of 29.6 grams of K_2SO_4 plus 0.3 gram NaCl in the same gaseous environment. This environment was selected because it borders test results with high and low corrosion rates and, accordingly, it would be more likely than most of the other enviroments investigated to induce SCC. Again, none of the specimens developed stress corrosion cracks, although all were somewhat tarnished on surfaces that contacted the salt.

Additional stressed C-ring specimens were exposed to the simulated incinerator atmosphere (as above) but in the absence of contact with either salt mixture or incinerator deposit. Neither cracking nor much corrosion attack resulted from these exposures. Metallographic examination of specimens exposed to hot salts and incinerator deposits revealed some instances of surface layer intergranular attack, but no evidence of SCC. In summary, SCC is not likely to be a problem with stainless steels at the operating temperatures of incinerators. However, the damp incinerator deposits can cause cracking.

LABORATORY CORROSION STUDIES

The objective of the laboratory experiments was to determine the importance of individual factors on corrosion in incinerators and also to help establish the mechanism by which metal wastage occurs. Since the experiments could be carried out under carefully controlled conditions, it was possible to determine the role of such factors as gaseous components, salts, or deposits, metal composition, and temperatures. The approach taken was to subject alloys to simulated flue gases with and without direct contact with known salt mixtures. Weight losses were used to assess the amount of corrosion which occurred.

Gas-Phase Reactions

As might have been expected, the attack of metals by gas-phase reactions was not great at normal operating temperatures, i.e., 800°F. Details of the extent of the attack from furnace gases containing 200 ppm HCl on A106 and T-11 steel at 800°, 1000°, and 1200°F are summarized in Table 3. It can be seen that the corrosion rates increased rapidly as the temperatures increased, in a manner similar to that noted on many incinerator corrosion probes although the estimated wastage of these laboratory specimens was much less than for the probe specimen. The presence of HCl had little effect on these corrosion rates.

TABLE 3: GAS-PHASE CORROSION AT ELEVATED TEMPERATURES

Alloy	Temperature, °F	Corrosion Rate, mils per month
A106	800	0.9
A106	1000	8
A106	1200	36
T-11	800	0.8
T-11	1000	6
T-11	1200	29

Direct Contact with Sulfate-Chloride Mixtures at 1000°F

Experience in oil- and coal-fired boiler furnaces has amply demonstrated the role of chlorine in fuels as an accelerating agent in corrosion processes. Corrosion by mixed sulfate-chloride salts has been observed. However, most of the reported work had been done at temperatures above 1000°F and with alloy steels rather than with carbon steels. Most of the work at 1000°F was carried out with a base corrosion mixture of K_2SO_4 and Na_2SO_4 in the molar ratio of 3 to 1. Detailed conditions concerning the corrosion mixture composition and gas atmosphere along with wastage rates at 1000°F are summarized in Figure 7.2. The lower line on this figure also shows the results of the azide test for the presence of sulfide on the corroded metals.

FIGURE 7.2: LABORATORY CORROSION RESULTS AT 1000°F

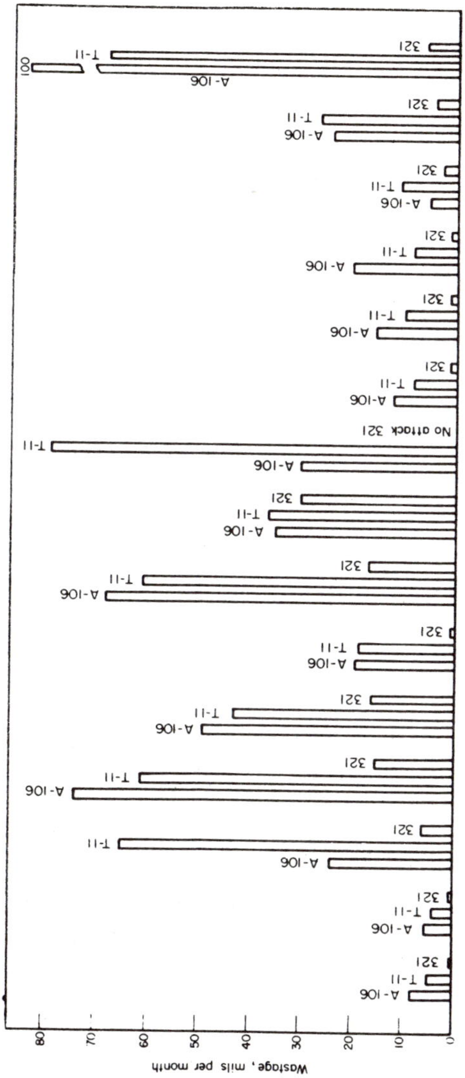

Run	0	3	1	5	20	7	10	11	13	15	16	18	17	19	21
K_2SO_4 %	-	59	58	78	78	99	75	75	-	-	78	78	78	78	75
Na_2SO_4 %	-	16	16	21	21	-	20	20	-	-	21	21	21	21	20
NaCl %	-	-	1	1	1	1	5	-	100	100	1	1	1	1	5
Misc %	-	25 Fe_2O_3	25 Fe_2O_3	-	-	-	-	5 KCl	-	-	-	-	-	-	-
Atmosphere	FG	FG	FG	FG	FG	FG	FG	FG	FG	No SO_2	No SO_2	No SO_2	Helium	He + SO_2	FG H_2SO_2
Azide test	-	?	++	+++	++	+	+++	+++	+++	-	-	-	-	-	+++

A triple +++ mark indicates a very strong test result. The corrosion is expressed in units of mils per month as calculated and extrapolated from the weight loss values for 50 hours of exposure. As can be seen in Figure 7.2, Run 3, the K_2SO_4 and Na_2SO_4 mixture with Fe_2O_3 added did not accelerate corrosion above that obtained in flue gas alone (Run 0). However, when 1 percent NaCl was added, the corrosion was greatly increased as indicated in Runs 1, 5, and 20. Runs 5 and 20 were repeat runs and demonstrate fairly good reproducibility.

The addition of Fe_2O_3 was made to determine whether reactions to form alkali iron trisulfates took place. Such compounds are known to be corrosive at 1100°F. Since the addition did not accelerate corrosion, it was concluded that the corrosiveness of the mixture of the possible formation of alkali-metal ferric trisulfates from sodium and/or potassium sulfates, iron oxide, and sulfur trioxide was not of major concern in the temperature range used.

As shown by Run 7, K_2SO_4 with 1 percent NaCl was less corrosive than the mixture of K_2SO_4 with NaCl. Increasing the chloride content had negligible effect although the action of KCl on A106 and T-11 steels at 5 percent level seemed to be somewhat less than for NaCl (Runs 10, 11).

It is significant that sodium chloride alone in the presence of flue gas was quite corrosive to the carbon steels, i.e., Run 13. Of even greater significance is the fact that sulfide as revealed by the azide test, was found on these specimens. Thus, some reaction involving the SO_2 in the flue gas was of importance since that was the only source of sulfur. Sulfide had been detected, of course, in all preceding runs in which corrosion had occurred. There was also a correlation between the severity of corrosion and the amount of sulfide detected.

In order to get a better understanding of the possible reactions involving sulfur and SO_2, several experiments were conducted with SO_2 eliminated. In addition, experiment No. 21 was carried out where the SO_2 content of the flue gas was increased from 250 to 2,500 ppm.

As can be seen from Runs 15, 16, and 18, the elimination of SO_2 from the flue gas decreased markedly the corrosion for chloride alone and for sulfate-chloride mixtures. Runs 16 and 18 were repeat experiments and it can be seen that results are in quite good agreement. It is important to note that no sulfide was detected on the metal surface of the specimens for the three runs just mentioned, i.e., with SO_2 absent. Exposure time was 50 hr in all runs.

Similarly the corrosion was low and no sulfide was formed when the exposures were carried out under helium with no SO_2 present (Run 17). The addition of SO_2 (2,500 ppm) to the helium increased the corrosion several times as indicated for Run 19. No sulfide was detected on these specimens.

In the experiment with the synthetic flue gas containing 2,500 ppm SO_2 (Run 21), a slight increase in corrosion rate was measured on the A106 steel, but the corrosion rates on the T-11 steel specimens were comparable to those when the level of SO_2 in the gas was 250 ppm (Run 10).

In other experiments it was found that extremely rapid attack of Type 321 stainless steel resulted from lead chloride at temperatures near 1000°F. Metallographic studies revealed intergranular penetration in addition to the general attack. The alloy was also attacked by molten zinc and zinc chloride vapor. The A106-Grade B and low alloy A213-Grade T-11 carbon steels were attacked about five times more rapidly at 1000°F when lead chloride was present than when it was absent.

Direct Contact Corrosion at 800°F

Seven experiments were carried out at 800°F under conditions similar to those discussed in the series at 1000°F. Results are summarized in Figure 7.3. In general, the accelerating effect of NaCl can be seen although wastage rates are much lower than at 1000°F. The effect of varying the SO_2 was not explored at this temperature.

FIGURE 7.3: LABORATORY CORROSION RESULTS AT 800°F

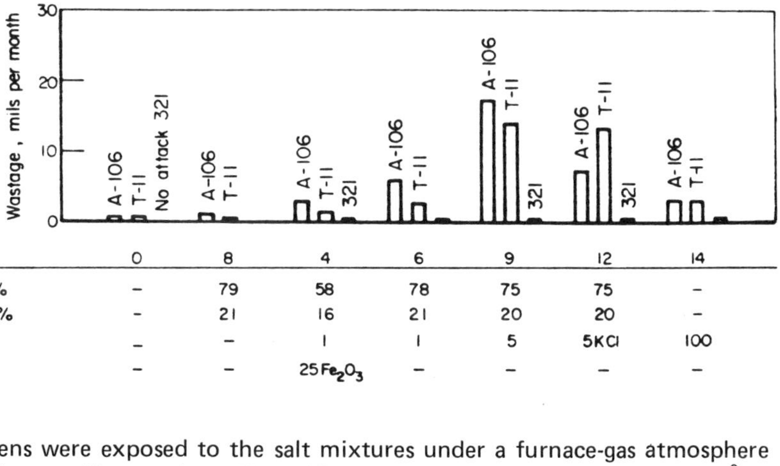

Run	0	8	4	6	9	12	14
K_2SO_4 %	-	79	58	78	75	75	-
Na_2SO_4 %	-	21	16	21	20	20	-
NaCl %	-	-	1	1	5	5KCl	100
Misc. %	-	-	25Fe_2O_3	-	-	-	-

Specimens were exposed to the salt mixtures under a furnace-gas atmosphere for 50 hours. No attack on Type 321 stainless steel was observed at 800°F, although this material had corroded in some experiments performed at 1000°F. The effects of alkali-metal chloride addition and of the presence of SO_2 in the flue gas at 800°F were similar, in general, to those observed at 1000°F with carbon steels. The corrosive action of KCl and NaCl at the 5 percent level was less at 800°F than at 1000°F.

Direct Contact Corrosion at 600°F

Since the field studies showed that fireside wastage could occur at metal temperatures of 600°F and lower, extensive laboratory studies were carried out with a variety of salt compositions. The standard synthetic flue gas was used, with additions of HCl as noted. The most corrosive salts examined at 600°F were the bisulfates and pyrosulfates. These were picked because of their low melting points and because they had been demonstrated to be of importance in low-temperature boiler-tube corrosion in fossil-fuel-fired boilers. They were also likely suspects here because the field studies showed that sulfur-bearing salts were involved in the corrosion.

The results with potassium bisulfate mixtures are summarized in the bar graphs in Figure 7.4. It can be seen in Run 48 that appreciable corrosion occurred in the pure salt and that the stainless steel was more severely attacked than were the carbon steels. The reason for this attack on the stainless steels is not apparent.

FIGURE 7.4: EFFECT OF $KHSO_4$ ON CORROSION AT 600°F

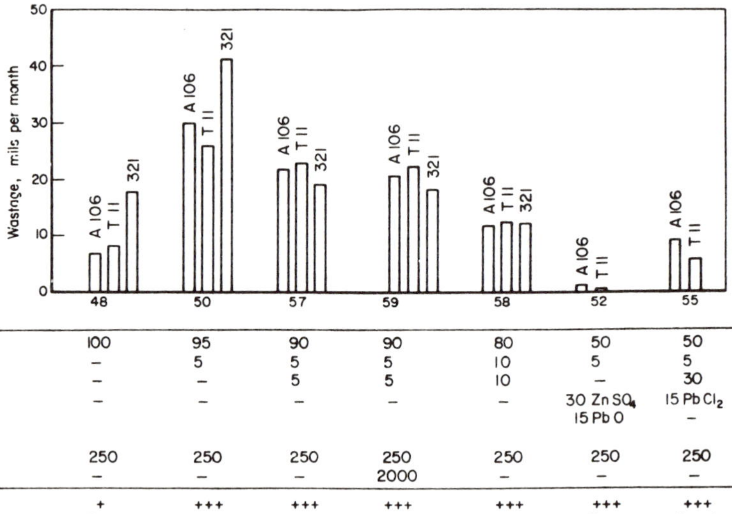

Run	48	50	57	59	58	52	55
$KHSO_4$ %	100	95	90	90	80	50	50
NaCl %	–	5	5	5	10	5	5
$ZnCl_2$ %	–	–	5	5	10	–	30
Misc %	–	–	–	–	–	30 $ZnSO_4$	15 $PbCl_2$
Misc %						15 PbO	–
SO_2 ppm	250	250	250	250	250	250	250
HCl ppm	–	–	–	2000	–	–	–
Azide	+	+++	+++	+++	+++	+++	+++
Physical state	Fused	Fused	Fused	Fused	Fused	Fused	Part fused

The addition of 5 percent NaCl to the bisulfate increased the attack by a factor of 3 or 4 (Run 50). The addition of 5 and 10 percent $ZnCl_2$ along with the NaCl did not increase corrosion (Runs 57 and 58). Similarly, the introduction of 2,000 ppm HCl to the flue gas did not accelerate corrosion (Run 59). The addition of $ZnSO_4$ and PbO to the NaCl-bisulfate mixture decreased corrosion (Run 52) and so did the addition of $PbCl_2$ and $ZnCl_2$ (Run 55). It should be pointed out, however, that at 1000°F, $PbCl_2$ and $ZnCl_2$ accelerate corrosion.

It should be noted that all these mixtures were molten during tests and that the corroded metals gave a positive test for sulfide.

As can be seen in Figure 7.5, $K_2S_2O_7$ caused greatly accelerated attack on carbon and stainless steels at 600°F (Run 28); however, dilution with inert components greatly reduced the action. The addition of NaCl caused very little increase in the attack (Run 44). The high rate of the attack on stainless steel was maintained when $ZnCl_2$ was added, but the attack on the carbon steels was decreased (Runs 36 and 41). Other additions of oxides, chlorides, and sulfates decreased the corrosiveness of the $K_2S_2O_7$ (Runs 46, 40, 30, and 37). Figure 7.5 shows that the more corrosive mixtures were those which had been molten during the test.

The mixture of K_2SO_4 and Na_2SO_4 and 5 percent NaCl which corroded the steels at 800° and 1000°F did not cause much attack at 600°F in 50 hours, as can be seen in Figure 7.5 (Run 22).

FIGURE 7.5: EFFECT OF $K_2S_2O_7$ ON CORROSION AT 600°F

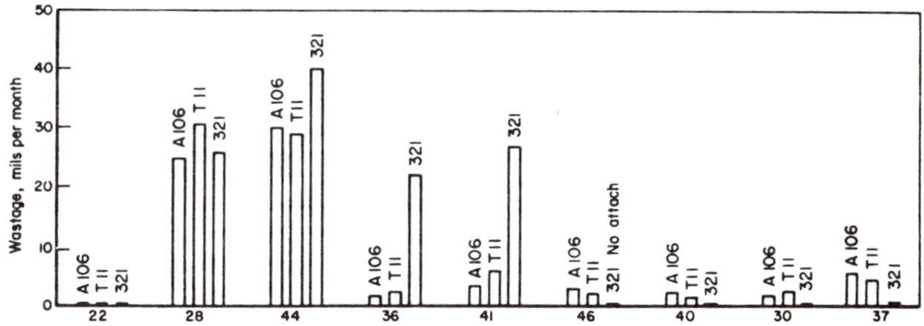

Run	22	28	44	36	41	46	40	30	37
$K_2S_2O_7$ %	–	100	25	95	95	65	50	5	5
K_2SO_4 %	75	–	–	–	–	–	–	71	67
Na_2SO_4 %	25	–	–	–	–	–	–	19	19
NaCl %	5	–	5	–	–	5	ZnO,CaO	5	5
Misc %	–	–	–	5 $ZnCl_2$	5 $ZnCl_2$	30 Fe_2O_3	Al_2O_3	–	5 $ZnCl_2$
SO_2 ppm	250	250	250	250	250	250	250	250	250
HCl ppm	–	–	–	–	–	–	2000	–	2000
Azide test	+	+++	+++	+++	+++	+++	+++	+	+++
Physical state	Agglom	Fused	Fused	Fused	Fused	Agglom	Caked	Caked	Caked

DISCUSSION OF CORROSION MECHANISMS

The results of studies of the deposits and scale formed on fireside tubes and probes in incinerators combined with the laboratory studies have provided an explanation of the cause of corrosion. Other discussions on the same subject can be found. As would be expected, several reactions are coupled and interrelated so that the overall picture is necessarily complex. The effects of various constituents are discussed individually; they can act in combination in practice.

Effect of Chlorides and Metal Salts

It is significant that chlorides are present throughout the deposits and at the scale/metal interface. The fact that ferrous chloride ($FeCl_2$) has been identified in the scale at the interface layer over the entire temperature range studied is considered to be important. It is thought that this iron chloride corrosion product is initially formed from the reaction of iron with the hydrogen chloride or elemental chlorine released at the scale/metal interface.

The corrosive effects from HCl would not be expected to be severe below temperatures of about 600°F. Since $FeCl_2$ has been detected on corrosion-probe samples exposed at temperatures well below 600°F, it is believed that some other mechanism is operating in this temperature range.

Although elemental chlorine has not been identified in incinerator furnace gases and would not be expected at high gas temperatures, it is believed that it may play a role in the corrosion reaction. It is postulated that metal oxides, possibly Fe_2O_3 or PbO, on the tube surfaces catalyze the reaction of hydrogen chloride with oxygen, to yield elemental chlorine and water. The chlorine, which is formed only near the catalytic surface, then combines directly with the iron.

It has been shown that once a layer of ferrous chloride has formed on the iron surface, reactions forming iron oxides (Fe_2O_3 and Fe_3O_4) and sometimes sulfides take place. Microscopic examination of the scales has shown that a molten salt phase has been present at points within the scale and as an interface layer between the scale and the deposit. This would act as a barrier to the motion of gases in and out of the layer adjacent to the metal surfaces.

Under oxidizing conditions, it would be expected that $FeCl_2$ (and also FeS) would react with oxygen to form Fe_2O_3. Under reducing conditions, on the other hand, it would be expected that this oxide would be reduced to Fe_3O_4. It is suggested that once a thin scale has formed on the metal tubes, such a cycling takes place. The mixed oxide and molten salt layers limit the availability of oxygen and retain chlorine and sulfur within the adherent scale. The chlorine retained within the scale deposit then reacts further with the iron to produce ferrous chloride.

Three observations suggest that chlorine may be involved. First, as was just mentioned, ferrous chloride is found next to the corroded tube wall even at the low temperature end of the corrosion probe. Second, the reaction of chlorine with carbon steel is extremely rapid at temperatures of 400°F and higher.

The curve in Figure 7.6 shows the high corrosion rate of carbon steel in dry chlorine at low temperatures. Third, the reaction of hydrogen chloride and oxygen, yielding free chlorine, is favored at temperatures near 500° to 700°F in the presence of catalysts. Thus it is suggested that chlorine reactions with steel are important at the low temperatures, and that, possibly, HCl and Cl_2

reactions take place at the high temperatures. The release of Cl_2 and HCl directly adjacent to the tube wall is an important part of the mechanism just proposed. It is of course necessary to explain how the chlorides reach the tube surfaces. It is suggested that the relatively short residence time in an incinerator would favor the deposition of any chloride originally present in the refuse. It has been well demonstrated that solid phase chlorides can be volatilized during burning processes. Chlorides observed in deposits can originate from HCl which is released when chlorinated plastics such as PVC are burned. This HCl then combines with the K_2O and Na_2O which are also refuse combustion products.

FIGURE 7.6: CORROSION OF CARBON STEEL IN CHLORINE AND HYDROGEN CHLORIDE

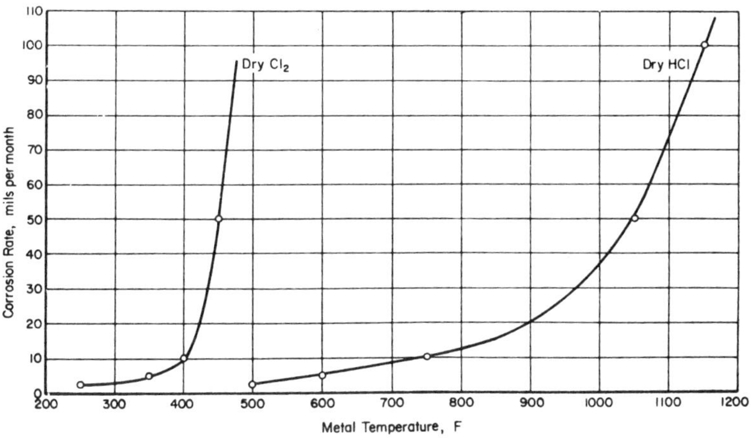

It is proposed that some of the chloride found on corroded surfaces has been transferred as $PbCl_2$ or $ZnCl_2$. Compounds such as dibasic lead phosphate, dibasic lead stearate, lead sulfate, lead chlorosilicate, dibasic lead phthalate, zinc 2-ethylhexoate, zinc laurate, and zinc stearate are added to PVC as thermal stabilizers. They act in this capacity because of their ability to react with the HCl liberated from the polymer as it is thermally decomposed. This can account at least in part for the presence of lead and zinc salts on the corrosion probes tested. Additional lead and zinc compounds of course could come from the volatilization of metal scrap in the refuse. They are probably first released as oxides in the flame and then are converted to sulfates and chlorides on the tubes.

Part of the deleterious function of the lead and zinc salts, particularly chlorides, could be the formation of molten salt layers, since their melting points are low and since when mixed with other chlorides (NaCl, KCl) they form eutectic compositions which melt at even lower temperatures. These low melting materials permeate and destroy the initial oxide scale so that the substrate metal becomes exposed to chlorine as it is released from the chlorides in subsequent reaction with furnace gases.

Effect of Sulfur Compounds

It is believed that sulfur bearing compounds play an important role with regard to corrosion at both low and high temperatures. For example, the most corrosive salts studied in the laboratory at 600°F were the pyrosulfates and the bisulfates. It is suggested that this high corrosivity is related to the fact that these materials have relatively low melting points, i.e., about 575°F for $K_2S_2O_7$ and 415°F for $KHSO_4$. Since the corrosion reactions, which are electrochemical in nature, can take place more readily in liquids than in solids or gases, the presence of a liquid phase is of great importance. It is believed, as was implied earlier, that the melting points of these salts is further modified, possibly lowered, by the zinc and lead salts also shown to be present on the corroded samples.

In addition to laboratory evidence showing the corrosivity of these materials, it should be pointed out that the azide and pH tests made on individual corrosion probe specimens show the presence of sulfide and the presence of acidic hydrolyzing salts on the majority of the specimens. Sulfur was also detected with the microprobe at areas near the metal surface.

CONCLUSIONS CONCERNING INCINERATOR WATERWALL CORROSION

General

Field and laboratory studies have demonstrated that the wastage in waterwall refuse boilers can be more severe than that normally encountered in fossil fuel fired boilers. The complex nature of the refuse used as the fuel and the relatively poorer control of burning in an incinerator combine to increase the possibility for corrosion. The contributors to the attack are corrosive gases and low-melting chloride and sulfur-containing salts which exert a fluxing action on the protective films on the metal surface.

These low-melting salts primarily contain compounds such as zinc and lead chlorides along with potassium bisulfate and potassium pyrosulfate. The data developed reveal that the gases SO_2, SO_3, HCl, and Cl_2 are also playing a major role in the wastage processes. Analyses of tube deposits and furnace gases confirm the belief that sufficient quantities of the deleterious salts and gases are present in all municipal incinerators to warrant careful consideration from a corrosion standpoint.

Tube Wastage

Work carried out with corrosion probes inserted in incinerators has demonstrated that the wastage rates of boiler tube metals are directly related to the operating metal temperatures. Some attack will be experienced in the 300° to 600°F range and greatly increased rates can be anticipated as temperatures are raised to 1000°F. The temperature gradient between the metal surface and the furnace atmosphere is also important. Thus, relatively low metal temperatures are to be preferred in flame areas.

The corrosion rates shown on the bar graphs earlier should not be used to project wastage values to be expected after long exposure periods such as a year or more. This is because the initial rates are always high and the rates decrease appreciably as the exposure time increases. The protective action of the tightly adhering scale probably accounts for this decrease in attack. Thus, rates for carbon steel tubes at about 500°F might be expected to range below about 10 mils per year for extended exposure periods rather than the much higher rates calculated using the bar graph data. However, scale removal to improve heat transfer could adversely affect the corrosion behavior.

While the stainless steels, particularly Types 310 and 446, furnish good resistance to fireside corrosion, their use is limited because of the deleterious effects which can take place during downtime under humid conditions. The Type 310 can fail because of stress corrosion cracking and the Type 446 can be severely pitted. However, only minor pitting occurred in incinerator operation. The other austenitic stainless steels evaluated, i.e., Types 304, 316, and 321, also were more resistant than the carbon steels but are subject to SCC. The Type 416 stainless steel did not perform well enough to be seriously considered. The Inconel 600 and 601 materials were quite resistant over the lower temperature region 300° to 500°F but were severely attacked at higher temperatures and thus are not good choices.

Incoloy 825 gave encouraging results over the entire temperature range and is much more resistant to SCC than the austenitic stainless steels so it could be considered for use in high temperature areas where the carbon steels would not be recommended. While the aluminum, chromium, and inorganic coatings furnished some initial protection to the tube surfaces they do not appear to be sufficiently durable to be considered for long time boiler operation.

Incinerator Operation

Several criteria of boiler operation can be suggested from the research results which should increase incinerator boiler life and reduce maintenance costs.

(1) Waterwall incinerators should not be used to generate high temperature superheated steam, but should be operated at relatively low metal temperatures, near 500°F, to minimize tube wastage. If high temperature superheated steam is desired for more efficient power production, it appears that there is merit in using the refuse to heat the water in a separate furnace and do the additional heating in another furnace with fossil fuel as the heat source. Thus corrosive deposits and gases will be kept away from tubes operating at high metal tem-

peratures. It is of interest in this connection to compare the nominal operating conditions of some of the waterwall installations on this continent with those in Europe and England. Given in Table 4 are steam temperatures and pressures and the nominal tube metal temperature, arbitrarily chosen to be 50 degrees above the steam temperature, for several stations operating and several being constructed.

It can be noted that the European practice, except for those at Rotterdam and Coventry, is to operate at relatively high temperatures and pressures. Severe metal wastage has been reported for some of these stations. On the other hand, the units at Norfolk have not shown corrosion. It is believed that the excellent performance of the Norfolk installation to date is related to the fact that the operating pressure there has been kept very low, i.e., 175 to 200 psi. Allowing for a 50°F rise through the metal wall, this corresponds to a metal surface temperature of about 425° to 440°F. Since the cool water flows to the bottom of the wall tubes and then rises, the metal temperatures in the flame area are probably even lower than 425°F.

TABLE 4: NOMINAL OPERATING CONDITIONS OF WATERWALL INCINERATORS

Location	Steam Pressure, psig	Steam Temperature, °F	Metal Temperature, °F (approx.)
Milan, Italy	500	840	890
Mannheim, Germany	1,800	980	1030
Franfurt, Germany	960	930	980
Munster, Germany	1,100	980	1030
Moulineaux, France	930	770	620
Essen Karnap, Germany	-	930	980
Stuttgart, Germany	1,100	980	1030
Munich, Germany	2,650	1000	1050
Rotterdam, Netherlands	400	680	730
Edmonton, England	625	850	900
Coventry, England	275	415	465
Amsterdam, Netherlands	600	770	820
Montreal, Canada	225	395	445
Chicago (N.W.), Illinois	265	410	460
Oceanside, New York	460	465	515
Norfolk, Virginia	175	375	425
Braintree, Massachusetts	265	410	460
Harrisburg, Pennsylvania	275	460	510
Hamilton, Ontario	250	400	450

It is interesting to note that the trend in design of new waterwall incinerators on this continent has been in the direction of operation at moderate boiler tube temperatures.

(2) It is concluded that downtime could be a critical corrosion period for a refuse boiler in that hygroscopic salts on the tubes could become wet and cause

attack. Such attack, however, has not occurred at Norfolk where the two boilers are operated alternately every other week. Nevertheless, it is believed that maintaining warm, dry tubes at all times is worthwhile. As a corollary it would appear that operation of the units 7 days a week, 24 hours a day would be optimum. Furthermore, metal temperatures in the boiler and in the effluent gas stream should be kept high enough to avoid either H_2SO_4 condensation or HCl/H_2O dew point condensation. Staying above about 300°F should present the former and above 225°F for the latter.

(3) Although not examined in the current program, the European practice of attempting to avoid all operation where reducing conditions can occur has much merit. In this connection the control of the burning and the feeding of the refuse can be of great importance. It should be pointed out that many years of research were required to develop the best techniques for burning the fossil fuels coal and oil with a minimum of corrosion. Fuel preparation and burning control are an important part of that technology. Similarly, if refuse is to be considered a fuel it should be treated as such and not simply fired as it comes from the pickup trucks.

INCINERATOR WET SCRUBBER CORROSION

A study was conducted on corrosion problems associated with wet scrubbers used with municipal incinerators. This grew out of the observation that scrubber corrosion was as serious or even more so than fireside corrosion.

There is general agreement that the corrosion in wet scrubbers for incinerators is an extremely severe problem. The rapid attack of most metals and alloys is not too unexpected when the composition and complexity of the effluent gases and dusts coming from the burning chamber are considered. These gases include corrosive constituents such as HCl, SO_2, SO_3, HF, and organic acids.

Extensive measurements of incinerator effluent gases have been reported during the past few years. One of the earlier measurements at the East 73rd Street Incinerator in New York City indicated an HCl evolution of up to 8 lb/ton of refuse. This incinerator did not use a scrubber but did contain water spray chambers. Thus, some of the HCl was probably removed ahead of the stack where the measurements were made. Since well designed scrubbers will remove much more of the soluble gases, it would follow that a relatively small incinerator of say 200 T/day could be absorbing up to 1,600 lb/day of HCl in the scrubber water. Large amounts of sulfur containing acids would also be absorbed.

The presence and concentration of HCl are of particular concern because this material accelerates pitting of most construction materials. It also limits the use of austenitic stainless steels because of the possibility of stress corrosion cracking. The importance of HCl in causing severe corrosion of equipment has been well documented. It has been shown that even trace amounts are damaging.

There are many possible sources of HCl in an incinerator. Some, but not all, of this gas comes from burning plastics. It has been found at many locations that the pH of the scrubber water soon drops to a low value, i.e., about 2.0. This is particularly true if the water is recirculated as is the practice in many places. An example of such an operation is at Montgomery County, Ohio. Data furnished by that office have been tabulated in Figure 7.7, which shows the buildup of various constituents in the scrubber water at the South Plant over about a 2 week period.

It can be seen that the pH remained quite low near 2.5 and that the chloride rose from about 9,000 ppm to 40,000 ppm. The sulfate, on the other hand, remained near 2,000 to 3,000 ppm and then abruptly rose to 25,000 ppm. Other dissolved solids, conductivity, etc., also rose as the operation continued. Thus, with all these acidic components present, it is estimated that the buildup of chloride in the circulating solution in the example cited above amounted to about 500 lb/day.

Other factors leading to high corrosion in incinerator scrubber are:

(1) Elevated temperatures (140° to 190°F)

(2) Abundant oxygen supply, and

(3) High velocities.

There is general agreement that:

(1) Carbon steels, cast iron, brass and bronze, are rapidly corroded.

(2) Performance of stainless steels has been variable, ranging from little corrosion to severe wastage and pitting.

(3) The more resistant alloys are Hastelloy C, titanium, and Inconel 625. However, they are quite expensive.

(4) Rubber and plastic coatings have shown some encouraging results, particularly in piping application.

(5) Fiberglas-reinforced plastic pipe and structural components show good resistance to the corrosive solutions as does acid-brick construction.

(6) Both deposits and corrosion can be a problem on induced-draft fans.

(7) Corrosion can be a problem on stacks.

(8) All components and areas in the scrubber are susceptible to corrosion, with the mist eliminator area being the most susceptible to attack.

(9) Concrete supports and flumes can be severely corroded by the hot acid solutions.

FIGURE 7.7: CHANGE IN COMPOSITION OF SCRUBBER WATER DURING CONTINUOUS RECIRCULATION

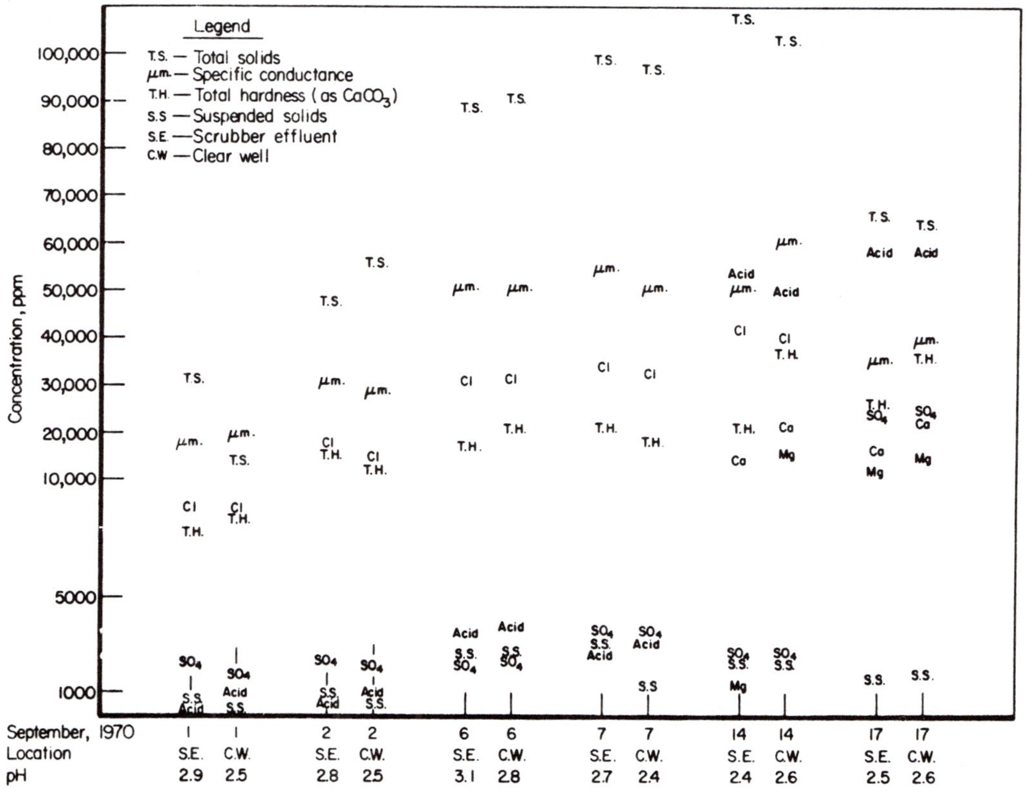

Scrubber Corrosion Studies

Procedures – A fairly extensive exposure program was carried out in the North Montgomery County Incinerator. The specimens were immersed in the hot water (170° to 190°F) passing through the concrete flume at the base of the scrubbers. The pH of this scrubber water varied from 4.0 to 5.1. A duplicate set of specimens was also exposed in the same water at Columbus, Ohio, under static, aerated, conditions at pH 2.0 to 2.5 to determine the effect of lower pH. Table 5 lists the eighteen metals and alloys evaluated in these scrubbers. Most were chosen because of their good corrosion resistance and resistance to stress corrosion cracking, although a few stainless steel compositions were included for comparison purposes.

TABLE 5: MATERIALS EVALUATED IN SCRUBBER

Stainless Steel	Hastelloy
304	C
310	C-276
316 L	F
446	G
Inconel	Carpenter
600	20 S
601	Type S-816
625	Titanium
Armco	75 A
22-13-S	6A1-4V
USS	
18-18-2	

A third set of stressed specimens of selected alloys was exposed at Columbus, Ohio, in contact with deposit removed from the ID fans and housing and maintained under hot, humid conditions above scrubber water.

Results from North Montgomery County Study – The specimens exposed in the scrubber water at North Montgomery County were examined after 8, 17, 38, 63, 96, and 129 days. These specimens rapidly became coated with a very adherent deposit which could be removed only by vigorous scrubbing with a bronze wire brush.

This coating very likely afforded some corrosion protection to the specimens. Spectrographic analyses of the coating showed a major amount of calcium and appreciable amounts of lead, barium, silicon, and aluminum. The low solubilities of calcium, barium, and lead sulfates is consistent with the presence of these deposits. X-ray diffraction studies confirm the presence of $PbSO_4$ and $CaSO_4{\cdot}2H_2O$.

A large portion of the water in this scrubber was recirculated, but a sizeable make up was required. This is evident from the data in Table 6 which includes analyses over a 4 month period. It can be seen that the chloride concentrations never exceeded 2,400 ppm, which is ten times lower than was reported for the South Plant where complete recirculation is employed. Furthermore, the results for 12/8/71 in Table 6 show extremely low concentrations of dissolved salts. This is because the scrubber during that period was being operated on a once-through water flow.

TABLE 6: SCRUBBER WATER COMPOSITION DURING EXPOSURE PERIOD

	Date			
Measurement	**8/31/71**	**10/8/71**	**11/2/71**	**12/8/71***
Total hardness as $CaCO_3$, ppm	2,330	2,760	6,250	272
Calcium as Ca, ppm	568	784	1,704	71
Magnesium as Mg, ppm	262	230	573	27
Sulfate as SO_4, ppm	1,261	1,308	1,933	231
Chloride as Cl, ppm	2,355	2,355	1,610	183
Specific conductance μmho/cm	7,499	7,639	11,160	833
Total solids, ppm	5,916	5,900	11,386	732
Suspended solids, ppm	114	14	10	114
Total acidity as $CaCO_3$, ppm	402	388	352	68
pH	4-4	4-3	5-1	4-0

*On this date the scrubber water was not being recirculated.

Details regarding each specimen exposed are found in Table 7 which lists the first instance of corrosion of stressed and welded specimens exposed to scrubber solutions and the severity of attack at the conclusion of the exposure. Corrosion rates based on weight loss measurements have not been presented because the attack when present was always localized in nature, so overall corrosion rates based on weight loss would be misleading.

The alloys are arranged with the most resistant ones at the top of each section (C-ring, U-bend, circular weld) and the least resistant at the bottom. Pit depths are shown for the plate specimens.

The most durable materials were Ti-6A1-4V, Hastelloy C, Inconel 625, Hastelloy G, and Hastelloy C-276, as shown by good ratings, in two groups of specimens. Some other alloys were rated good in one section but not in others. The least resistant were Type 304, 304 sensitized, Armco 22-13-5, USS 18-18-2, 445 and Inconel 600. The other materials were intermediate between these extremes. The unalloyed titanium (Ti 75A) was quite resistant and the weld cracks reported in the table may have been present in the as-welded specimens.

The only definite cracking as a result of exposure was on the Type 304, 304 sensitized, and the Inconel 600 specimens. In many respects the attack has the appearance of stress accelerated corrosion rather than stress corrosion cracking, particularly for the Inconel.

TABLE 7: RESULTS AT NORTH MONTGOMERY COUNTY, OHIO*

------Specimen------ Alloy	Number	Corrosion Results	Depth of Attack, mils Pit	Crevice
		U-Bends		
Ti-6Al-4V	J1	Good resistance 129 days		
	J2	Good resistance 129 days		
Hast C	D1	Good resistance 129 days		
	D	Good resistance 129 days		
625	N1	Good resistance 129 days		
	N2	Good resistance 129 days		
Hast G	S1	Good resistance 66 days		
(welded)	S2	Good resistance 66 days		
Hast C276	R1	Good resistance 66 days		
(welded)	R2	Good resistance 66 days		
316 L	H1	Good resistance, one pit 129 days		
	H2	Good resistance 129 days		
Carp 20	C1	Very fine pits 63 days, same at 129 days, good resistance		
	C2	Very fine pits 63 days, same at 129 days, good resistance		
Hast F	E1	Very fine pitting 63 days, slight edge attack 129 days		
	E2	Good resistance 129 days		
Armco 22-13-5	T1	Good resistance 33 days		
	T2	Good resistance 33 days		
T75A	K1	Weld-grains etched, possible cracks in weld, 129 days		
	K2	Weld-grains etched, possible cracks in weld, 129 days		
S816	F1	Slight roughening 63 days, same at 129 days		
	F2	Slight roughening 63 days, same at 129 days		
USS 18-18-2	U1	Bad edge and end-grain attack at 33 days		
	U2	Bad edge and end-grain attack at 33 days		
		C-Rings		
825	A1	Good resistance 129 days		
	A2	Good resistance 129 days		
601	M1	Good resistance, slight corrosion at machining lines 129 days		
	M2	Good resistance, slight corrosion at machining lines 129 days		
310	1G	One pit near crevice 63 days, slight pitting near crevice 129 days		
	1H	Slight pitting near crevice 63 days, same at 129 days		
316 L	H1	Some edge attack 63 days, worm track and edge attack 129 days		
	H2	Edge attack 63 days, same at 129 days		
304	3G	Pitting 17 days, rows of pits across face 63 days, possible cracks 129 days		
	3H	Pits 8 days, rows pits across face 63 days, possible cracks 129 days		
304 S	2G	Pitting 38 days, rows pits across face 63 days, possible cracks 129 days		
	2H	Pitting 38 days, rows pits across face 63 days, possible cracks 129 days		
600	B1	Pitting 17 days, deep grooves 63 days, same at 129 days		
	B2	Edge attack 8 days, deep grooves 63 days, same at 129 days		
446	4G	Pitting 17 days, severely corroded, one penetration 129 days		
	4H	Pitting 8 days, severely corroded, deep edge canyons 129 days		
		Circular Welded Plates		
Ti-6Al-4V	J	Good resistance, grains outlined 129 days	0	0
625	N	Good resistance 129 days	0	0
Hast C	D	Good resistance 129 days	0	0
Hast F	E	Good resistance 129 days	0	0
S816	F	Good resistance 129 days	0	0
Carp 20	C	Some crevice attack 129 days	0	4
825	A	Some crevice attack 129 days	0	5
Armco 22-13-5	T	Pitting and crevice attack 33 days	2	15
316 L	H	Pitting and crevice attack 63 days, worse at 129 days	8	20
310	I	Pitting 38 days, deep pits and crevice attack at 129 days	14	32
600	B	Shallow pits 8 days, bad attack surface, weld. edge, crevice at 129 days	30	28
304	G	Pitting 8 days, deep pits, more at heat-affected zone 129 days	20	31

* pH 3.0 to 5.1. Temperature 170° to 180°F. Most specimens examined at 8, 17, 63, 96, and 129 days.

Cracking was not observed in the circular welded panels, but many of them were pitted, often predominantly in the area near the weld. The strings of pits may be similar to those which finally produced leakage in the actual scrubber wall. Similar selective attack at the support hole crevices and the pits in the weld area on the back of the Type 310 stainless steel panel were evident.

Type 304 stainless steel was similarly affected, except that more linear strings of pits were present. Inconel 600 was usually severely attacked at corners and edges and crevices. It will be recalled, however, that stressed C-rings of this alloy showed deep grooves.

Results from Columbus, Ohio, Study – The scrubber solutions used in the exposures made at Columbus, Ohio, were obtained from the scrubber effluent stream at the North Montgomery County Incinerator. The solutions were changed three times during the 137 day period. A set of metal specimens identical to those exposed at North Montgomery County was used.

The specimens were completely submerged in the scrubber water contained in a large Pyrex jar maintained at 170° to 180°F. Air was slowly sparged through the solution during the exposure. The pH of the solutions was maintained at 2.0 to 2.5 by means of occasional additions of sulfuric or hydrochloric acids.

Specimens were examined at 20, 42, 95, and 137 days. The results in Table 8 have been arranged and tabulated in a manner similar to that used for the specimens described in the preceding section. The following materials furnished the best resistance when rated on the basis of good performance in all configurations in which they were run: Ti-6Al-4V, Hastelloy C, Inconel 625, Alloy S816, and Hastelloy F, Hastelloy G, and Hastelloy C-276.

It will be noted that stress corrosion cracking was found on the Type 304 panels and C-rings. Cracking was also noted at an area in the weld and base metal of the Type 316 L panel.

Stress accelerated corrosion (trenches) running across the face of the specimen was found for the Inconel 600 and 601 specimens. Pitting occurred on many of the stressed and welded specimens exposed to scrubber solutions as noted in Table 8. It will be noted in Table 8 that most of the panels which showed bad pitting were also severely attacked at the crevices under the Teflon spacers near the support holes in the corners of the panels.

Fan Deposit Studies at Columbus, Ohio – There have been field service failures of induced draft fans, particularly from SCC when austenitic stainless steels were used. Carbon steel fans have also corroded rapidly by general attack and by pitting.

This subject has been explored experimentally by subjecting C-rings and U-bends of various alloys to deposits taken from the fan housing at the incinerator at North Montgomery County.

TABLE 8: CORROSION RESULTS AT COLUMBUS, OHIO*

Specimen Alloy	Specimen Number	Corrosion Results	Depth of Attack, mils Pit	Depth of Attack, mils Crevice
		U-Bends		
Ti-6Al-4V	J3	Good resistance, 137 days		
	J4	Good resistance, 137 days		
Hast C	D3	Good resistance, 137 days		
	D4	Good resistance, 137 days		
625	N3	Good resistance, 137 days		
	N4	Good resistance, 137 days		
Hast F	E3	Good resistance, 137 days		
	E4	Good resistance, 137 days		
Carp 20	C3	Good resistance, 137 days		
	C4	Good resistance, 137 days		
S816	F3	Good resistance, 137 days; surface mottled		
	F4	Good resistance, 137 days; surface mottled		
316 L	H3	Slight stringer corrosion at edges, 41 days, same 137 days		
	H4	Good resistance, 137 days		
Hast G	S1	Good resistance, 66 days		
(welded)	S2	Good resistance, 66 days		
Hast C276	R1	Good resistance, 66 days		
(welded)	R2	Good resistance, 66 days		
Ti75A	K3	Weld grains outlined 42 days; same 137 days, fine weld cracks		
	K4	Weld grains outlined 42 days; same 137 days, fine weld cracks		
Armco 22-13-5	T3	Edge pits, 33 days, same at 66 days		
	T4	Edge pits, 33 days, same at 66 days		
USS 18-18-2	U3	Edge pits, 33 days, worse at 66 days, surface roughening		
	U4	Edge pits 33 days, worse at 66 days		
		C-Rings		
825	A3	Good resistance, 137 days		
	A4	Good resistance, 137 days		
310	1J	Good resistance, 137 days		
	1K	Good resistance, 137 days		
316 L	H3	Good resistance, 137 days		
	H4	Good resistance, 137 days		
600	B3	Fine pits, 42 days; trenches, pits across face, edge pits, 137 days		
	B4	Fine pits, 42 days; trenches, pits across face, edge pits, 137 days		
446	4J	Pitting at 20 days, more pitting at 137 days		
	4K	Pitting at 20 days, more pitting at 137 days		
601	M3	Bad edge pitting, 43 days, deep trenches and pits across face, 137 days		
	M4	Edge pitting, 95 days, trench formation, pitting, 137 days		
304	3J	One crack at 20 days, deep crack at 137 days, some small cracks		
	3K	Possible crack at pitted area, 137 days		
304	2J	Fine cracks at 95 days, same at 137 days		
Sens	2K	Three cracks at 20 days, complete penetration at one crack, 137 days		
		Circular Welded Plates		
Ti-6Al-4V	J	Good resistance, 137 days; weld grains outlined	0	0
Hast C	D	Good resistance, 137 days	0	0
S816	F	Good resistance, 137 days; surface markings	0	0
625	N	Good resistance, 137 days	0	0
Hast F	E	Good resistance, 137 days	0	0
600	B	Roughened, pitted at 95 days; same at 137 days	3	0
Carp 20	C	Pits near weld, attack at crevice 93 days; same at 137 days	3	20
310	I	Pitted on back near weld, 42 days; attack at crevice, 137 days	5	48
825	A	Pits in weld, 95 days; pits front and back, 137 days	8	3
316 L	H	Pitting on back welded area, 42 days; more pitting, 137 days crevice attack, crack in and near weld	7	12
Armco 22-13-5	T	Pits in and near weld, crevice attack, 33 days; same at 55 days	3	7
304	G	Crack through plate near weld, 42 days; two bad cracks, 137 days	0	5

*pH 2.0 to 2.5. Temperature 170° to 180°F. Aerated. Most specimens examined at 20, 42, 95, and 137 days.

Deposit Analyses

Deposits from both the blade and the housing of the induced draft fan at the Montgomery County North Incinerator were analyzed during studies. These deposits were substantially different from those found on the metal corrosion probes at the Miami County and Norfolk incinerators. The difference results primarily from the fact that a wet scrubber has removed most of the particulate material from the gas stream before it reaches the induced draft fan.

As a result, the amount of iron, aluminum, silicon, and calcium found in the fan deposit is low. These are the components of the clay-like materials in the fly ash which were removed in the scrubber. The lead, zinc, and potassium content was high, and the compounds $K_2Pb(SO_4)_2$ and $K_2Zn(SO_4)_2$ were readily identifiable in the deposits. The presence of these compounds in the fan deposits shows that some particles escape the scrubber water, as these sulfates are soluble but will precipitate as $PbSO_4$ when exposed to water.

The chlorine content of these deposits was unusually high. The chloride concentration was almost 16 percent in the housing deposit. This buildup of chloride in the ID fan deposits shows that the wet scrubber is not removing all the HCl vapors.

Corrosion

The stressed specimens were placed edge down and buried about one-half inch in the deposit, which was held several inches above a sample of scrubber solution. The entire assembly was contained in a closed vessel which opened at the top to take a reflux condenser. The scrubber solution was heated to 170° to 180°F to produce humid conditions within the container. Thus, the stressed samples touched damp deposit and vapors from the scrub water.

The results for the twelve alloys including carbon steel which were exposed up to six weeks are given in Table 9. Type 304 stainless steel and plain carbon steel were severely pitted after two weeks and were removed, while Armco 22-13-5 was severely cracked and removed at this time. The alloys have been arranged with the most resistant alloys at the top of the list and the least resistant at the bottom of the table.

The degree of rusting, pitting, and cracking is indicated by numbers from 0 to 9. Pit depths are also shown. The best resistance was furnished by Ti-6Al-4V, Inconel 625, and Hastelloy C. A small amount of pitting was observed on the Type S816, Incoloy 800, and Incoloy 825 specimens. Stress corrosion cracking and pitting was detected by surface examination of the Carpenter 20, Type 316 L, USS 18-18-2, and Armco 22-13-5 specimens.

It can be seen from the results just discussed that the corrosion under the humid fan deposits was more severe than in the scrubber water. This is undoubtedly related to the presence of localized high concentration of chloride salts at the metal surfaces and the absences of scale forming protective deposits.

TABLE 9: CORROSION RESULTS OF STRESSED SPECIMENS EXPOSED TO INDUCED DRAFT FAN HOUSING DEPOSITS UNDER HOT HUMID CONDITIONS

		Corrosion Found at Indicated Week*					
Alloy Type	Specimen	1	2	3	4	6	Pit Depth, mils
Ti-6Al-4V	J	G-0, P-0, C-0	G-0, P-0, C-0	G-0, P-0, C-0	G-0, P-0, C-0	G-1, P-0, C-0	0
Inconel 625	N	G-0, P-0, C-0	-	G-0, P-0, C-0	G-0, P-0, C-0	G-0, P-0, C-0	0
Hastelloy C	D	G-0, P-0, C-0	G-0, P-0, C-0	G-0, P-0, C-0	G-1, P-C, C-0	G-1, P-0, C-0	0
S816	F	G-0, P-1, C-0	-	G-0, P-1, C-0	G-1, P-2**, C-0	G-1, P-2**, C-0	2
Incoloy 800	W	G-1, P-1, C-0	-	G-1, P-1, C-0	G-1, P-1, D-2, C-0	G-1, P-2, D-3, C-0	6
Incoloy 825	A	G-0, P-2**, C-0	G-0, P-3**, C-0	G-0, P-3**, C-0	G-0, P-3**, C-0	G-0, P-4**, D-2**, C-0	15
Carpenter 20	C	G-0, P-2**, C-0	-	G-0, P-2**, C-0	G-0, P-2**, D-2**, C-1**	G-0, P-3**, D-3**, C-1**	9
316 L	H	G-1, P-4, D-2, C-0	G-1, P-4, D-2, C-0	G-2, P-4, D-3, C-1	G-2, P-4, D-3, C-1	G-2, P-4, D-4, C-2	16
USS 18-18-2	U	G-3, P-3, C-0	G-4, P-3, C-0	G-5, P-4, C-0	G-5, P-4, C-1	G-6, P-4, D-1, C-2	7
Armco 22-13-5	T	G-1, P-3, C-7	G-1, P-3, C-8	(removed)	-	-	8
304	G	G-5, P-5, C-0	G-8, P-5, C-0	(removed)	-	-	-
Plain Carbon Steel	V	G-8	G-9 (removed)	-	-	-	-

*Magnitude of corrosion increasing 0-9 by surface examination.
G = General rust
P = Surface pitting
C = Cracking

**Major attack where specimens contact deposits.

The most severe surface cracking was observed on Armco 22-13-5. More rusting and pitting took place on the USS 18-18-2 alloy, although cracks were seen in the face view at 7X. The examination of cross section was required in some cases to reveal the SCC. This was either because of the presence of severe rusting, which masked the cracks as for Type 304, or because the cracks were small, as for Type 316 L and Carpenter 20.

It will be noted that the depth of the cracks in the alloy just mentioned when viewed in cross section was about 40 mils. Similar penetration of stress corrosion cracks were observed with Type 316 L stainless steel. However, in this case less surface rusting occurred. The depth of the attack on the Carpenter 20 specimen was only about 5 mils.

Conclusions from Scrubber Studies

The studies of alloys under three scrubber exposure conditions have furnished useful information on their general corrosion and stress corrosion cracking behavior. The superior resistivities of Ti-6Al-4V, Inconel 625, and Hastelloy C as reported by other investigators (such as the International Nickel Company) have been confirmed. Unalloyed Titanium 75A may also be good, but since some fine cracks in the weld were visible it has been rated lower. It is possible that these cracks were present in the as-welded material.

Hastelloy F and S816 have also shown good resistance, as have Hastelloy G (welded) and Hastelloy C-276 (welded). The last two alloys were exposed for only 66 days, however. The studies indicate that the alloys mentioned above are also resistant to stress corrosion cracking under the conditions used.

Conditions present when metals are in contact with ID fan deposits are more likely to produce stress corrosion cracking than are the conditions present in scrubber waters. Thus, the stainless steels, Types 304, 316 L, USS 18-18-2, Armco 22-13-5, and Carpenter 20, were cracked to varying degrees when in contact with these deposits.

In scrubber solutions, on the other hand, Types 304 and 316 L were the only materials that cracked. This was true for both sensitized and unsensitized Type 304 material. It is significant, however, that stress apparently had a marked effect on the corrosion of Inconel 600 and 601 in scrubber solutions because deep grooves were produced in the stressed area.

Pitting and crevice attack seemed to be a more severe problem with many alloys in the hot scrubber solutions. For example, pitting was noted on Type 304, 310, 316 L, 446, Carpenter 20, USS 18-18-2, Armco 22-13-5, and Incoloy 825. The leakage seen in some scrubbers constructed of 316 L stainless steel reflects the pitting tendency of this alloy in scrubber environments.

No preferential attack was noted at crevices where the C-rings or U-bends touched. However, severe attack as noted in the tables was found with many alloys at areas under the Teflon spacers near the corner holes in the 4 x 4 inch panels. Because of the cost of the more corrosion resistant materials for incinerator scrubber service, it appears that consideration should be given to the use of nonmetallic construction materials. The good performance of such materials at the East 73rd Street Incinerator also is an important consideration.

PART II.
SPECIAL INCINERATOR APPLICATIONS

Incineration of Sewage Sludge

The material in this chapter is based upon PB 197,888 April 1970, prepared by S. Balakrishnan, D.E. Williamson and R.W. Okey of Resource Engineering Associates; EPA *PB 211,323* August 1972; and the EPA *News of Environmental Research in Cincinnati,* May 3, 1974 by R.A. Olexsey and J.B. Farrell.

Primary sewage treatment is specifically designed to remove the bulk of the solids present in the sewage. The removed solids, in the form of sludge, are either disposed of directly from the primary process or receive their final disposal after combination with sludges from secondary treatment processes.

In certain areas, particularly on the East Coast, the practice has been to barge sewage sludge to sea, for ultimate disposal into the ocean depths. It is this form of disposal which has, over the past year, been the subject of official concern and action because of its apparent adverse effect upon life on the ocean floor. In October 1970, the Council on Environmental Quality recommended, in its *Ocean Dumping—A National Policy,* that ocean dumping of sludge should be phased out as an ultimate disposal practice. The Marine Protection, Research and Sanctuaries Act of 1972 is designed to eventually eliminate ocean dumping. Incineration will become the likely choice for sludge disposal by communities presently barging their sludge to sea if ocean dumping is prohibited outright.

The cost of the conventional sludge digestion system constitutes a significant portion of the total cost of treatment plants and yet sludge digestion does not provide for the maximum reduction nor the ultimate destruction of the remaining waste organic solids. Further, anaerobic reactors are extremely difficult to operate and frequently cause as many problems as all the rest of the plant combined. For these reasons, sludge disposal is becoming an important factor to be considered in the design of new plants and expansion of existing systems.

There are other pressures which are derived from our society and the way it is changing and growing. First, the quantity of sludge will increase by some 60 to 70% by 1980, due both to the increase in population and in the degree of treatment. Secondly, the factors mitigating against land disposal will increase because of mounting desires to avoid the indiscriminate disposal of waste due to aesthetic and health reasons. Thirdly, sludge handling by the older techniques frequently represents 20 to 40% of the capital and operating cost of the treatment plant. These methods were evolved when labor was cheap and the situation now is dramatically different. Now, more sophisticated operational techniques are being evolved.

The on-site disposal of sludge is compatible with all these driving forces and incineration seems to be the only practical means presently known that can accomplish maximum reduction of waste solids. It is expected that future improvements in the sludge combustion practices will reduce costs.

Incineration of sludge appears able to meet the ultimate goal of the efficient disposal of solid waste material without causing extensive nuisances to the community. The trend seems to be away from digestion on account of the increased capacity required, owing to increased use of detergents, and toward dewatering in vacuum filters or pressure filters and disposal of the sludge cake on land directly or after composting with ground-up refuse. There is increased interest in the country in combining incineration of sludge cake together with refuse.

Until 1935, no sludge incineration was practiced in the United States, but by 1972, the United States was incinerating 2,750 dry tons of sludge per day, or 25% of total raw sludge generation (4.2 million tons per year). The use of incineration as a sludge disposal technique is rapidly increasing, as shown by installation figures for the United States from four major manufacturers:

Year	No. of Units Installed	Year	No. of Units Installed
1963	18	1968	28
1964	16	1969	27
1965	20	1970	26
1966	10	1971	44
1967	24		

New sludge incinerator installations are expected to average 70 units per year in the 1970s. Incineration is generally more costly than alternative methods such as sanitary landfill or landspreading. However, these alternative methods require large land areas if offensive conditions are to be avoided. Increased cost and reduced availability of land near urban wastewater treatment facilities have created the trend toward incineration, which provides a sanitary method for disposal of sludge in a relatively small land area.

CHARACTERISTICS OF SEWAGE SLUDGE

The composition of sewage sludges varies widely depending on a complexity of factors. Primary sludges are higher in caloric value than secondary (biological) sludges because of their high grease content. It is more economical to burn undigested solids then the digested solids since digestion significantly reduces the heat content of the remaining solids. The average characteristics of sewage solids are summarized in Table 1.

TABLE 1: AVERAGE CHARACTERISTICS OF SEWAGE SLUDGE

Material	Combustibles (%)	Ash (%)	Btu/lb
Grease and scum	88.5	11.5	16,750
Raw sewage solids	74.0	26.0	10,285
Fine screenings	86.4	13.6	8,990
Ground garbage	84.8	15.2	8,245
Digested sewage solids and ground garbage	49.6	50.4	8,020
Digested sludge	59.6	40.4	5,290
Grit	30.2	69.8	4,000

Source: PB 197,888 1970

Sludge ratios of primary to secondary are generally 8 to 10:1 to insure aerobic conditions in the thickener. The primary to secondary sludge ratio directly affects both dewaterability and heat value of the sludge. Experience shows that primary plus secondary sludge from trickling filters will produce about 7% solids in the thickener and dewater to about 25% solids, whereas primary and activated sludge will thicken to 5% solids and dewater to 22% solids. The difference in terms of auxiliary fuel costs could be between $2 and $3 per ton of dry solids.

Heat Value of Sewage Sludge

The combustible elements of sewage sludge are carbon, hydrogen and sulfur and these elements are chemically combined in the organic sludge as grease, carbohydrates and protein. The combustible portion of sewage sludge has a Btu content equal to that of lignite coal. Air is added to provide oxygen to support the combustion of the combustible elements. The reactions of these elements with oxygen are as given in Table 2.

The composition of elements in sewage sludge varies from plant to plant as shown in Table 3. The heat value of sewage sludge can be estimated if its ultimate analysis is known. DuLong's formula can be used to compute the heat value as shown on the following page.

$$Q = 14{,}600\ C + 62{,}000\ (H - \frac{O}{8})$$

Q = Btu/lb of dried sludge
C = % carbon
H = % hydrogen
O = % oxygen

A reduction in the thermal value of the sludge occurs when inorganic chemicals are added to aid filtration. These chemicals used are inert and, therefore, lower the heat content per pound of filter cake. The other disadvantage is that the weight of the sludge is increased by 10 to 15% by the addition of chemicals. The heat energy available for sustaining combustion will be reduced if ferric hydroxide and calcium hydroxide sludges are burned with the sewage sludge due to the heat used up in the dehydration of these hydrous sludges.

TABLE 2: COMBUSTION REACTIONS OF SEWAGE SLUDGE

Reaction	Heat Release (Btu/lb)
(1) Carbon + oxygen → carbon dioxide $C + O_2 \longrightarrow CO_2$ (1 lb) (2.67 lb) (3.67 lb)	14,500
(2) Hydrogen + oxygen → water $2H_2 + O_2 \longrightarrow 2H_2O$ (1 lb) (7.94 lb) (8.94 lb)	62,000
(3) Sulfur + oxygen → sulfur dioxide $S + O_2 \longrightarrow SO_2$ (1 lb) (1 lb) (2 lb)	4,500

Source: PB 197,888 1970

TABLE 3: ELEMENTAL COMPOSITION OF SEWAGE SLUDGE

	Source*			
Composition	No. 1	No. 2	No. 3	No. 4
Carbon (%)	64.3	65.6	55.0	51.8
Hydrogen (%)	8.2	9.0	7.4	7.2
Oxygen (%)	21.0	20.9	33.4	38.0
Nitrogen (%)	4.3	3.4	3.1	3.0
Sulfur (%)	2.2	1.1	1.1	Trace
Volatiles (%)	47.9	72.5	51.4	82.0

(continued)

TABLE 3: (continued)

Composition	Source* No. 1	No. 2	No. 3	No. 4
VSS (Btu/lb)	12,840	12,510	10,940	8,990
TSS (Btu/lb)	6,160	9,080	5,620	7,380

*Source No. 1 - Cleveland Southerly Plant, 1955
No. 2 - Detroit, Michigan, 1955-56
No. 3 - Minneapolis, Minnesota, 1955
No. 4 - New Rochelle, New York, 1960-62

Source: PB 197,888 1970

IMPROVEMENTS IN THE HEAT VALUE OF SLUDGE

Improvements in the heat value of sludge can be achieved only by improving the volatile content of sludge as fed to the combustion unit as no control can be exercised over the stoichiometry of sludge combustion. The volatile content of a given sludge may be improved by a very efficient degritting system. Hydrocyclones, used for this purpose, have shown removals of 95% of the plus 200 to 270 mesh inorganics at a specific gravity of 2.67 and increases in the volatile content of the sludge from 70 to 75% to 80 to 85%. The effect of volatile content on the operating cost of sludge combustion is shown in Figures 8.1 and 8.2.

FIGURE 8.1: EFFECT OF VOLATILES IN SLUDGE ON FUEL (NATURAL GAS) COST

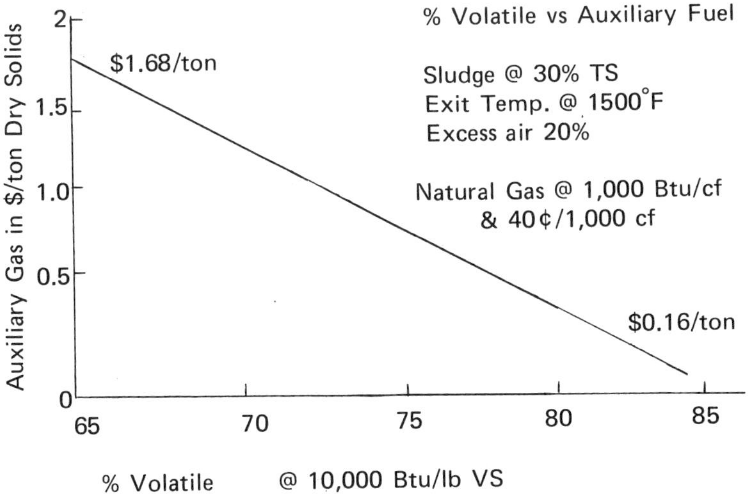

Source: PB 197,888 1970

FIGURE 8.2: EFFECT OF VOLATILES IN SLUDGE ON FUEL (NO. 2 OIL) COST

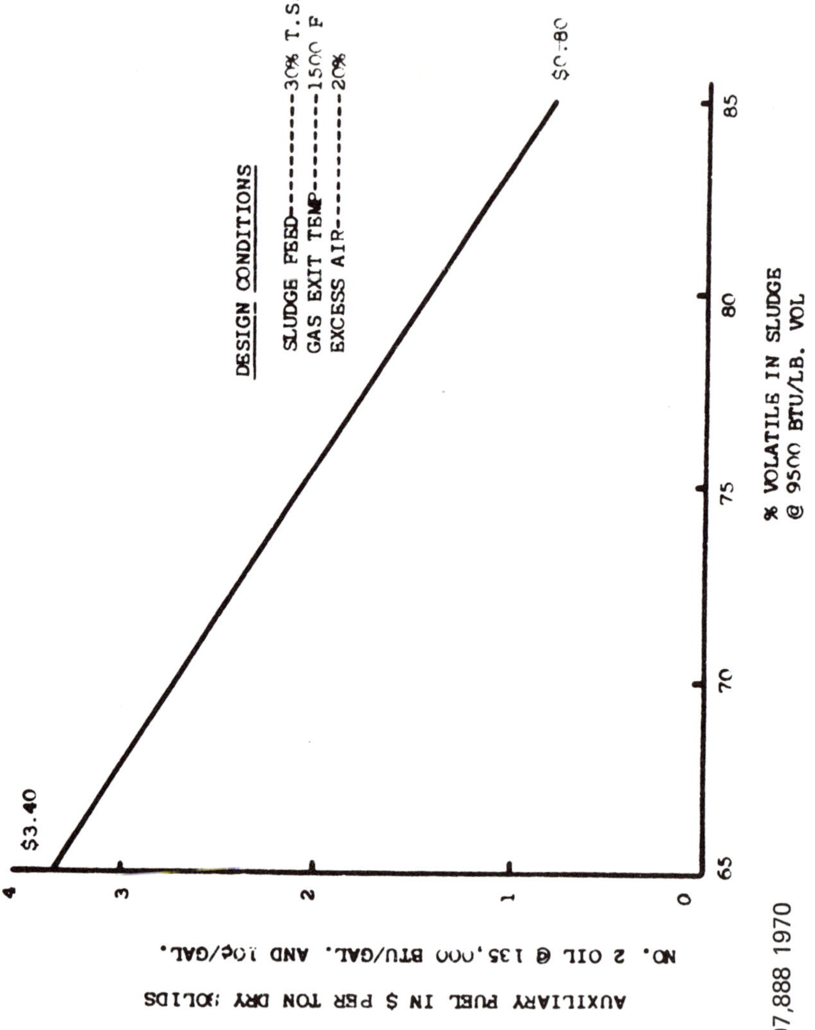

Source: PB 197,888 1970

A flocculation process used in conjunction with clarification in the primary treatment area, increases the sludge settling rates and, therefore, the ratio of primary to secondary sludge. Assuming an overall removal of 95% for conventional activated sludge, the sludge resulting from such a process will thicken (7%) and dewater (25%) to the same degree as primary plus trickling filter. Therefore, there is a saving on fuel and an increase in a combustion unit's capacity.

Consideration must also be given to materials present which will react endothermically at combustion temperatures. The moisture content of the calcium carbonate ($CaCO_3$) sludge and the endothermic decomposition to calcium oxide (CaO) materially increases the thermal burden. The off gas cleaning requirement would also be increased.

Further, a highly alkaline sludge would be more difficult to dispose of than the original calcium carbonate, except in some cases where liming of soils is required. Therefore, the calcium carbonate sludge should be handled separately and dewatered further, if necessary, for final disposal by landfill.

The following discussions relating to auxiliary fuel requirements, and to process variables, are largely specific for fluidized bed incinerators. However, the discussions broadly apply to other types of incinerators.

AUXILIARY FUEL REQUIREMENTS

The two important factors that affect the auxiliary fuel requirements are the heat value of the sludge and the heat required for adequate burning. By adequate burning, it is meant the heat required for complete incineration of the sludge and to raise the temperature of the gases to a sufficient level to insure odor control. The magnitude of the temperature requirement depends upon the nature of the sludge being burnt but the minimum deodorizing temperature for conventional incineration units has been established at 1350° to 1400°F.

The heat required for the incinerator system depends primarily on the efficiency of burning and the degree of excess air required. The following constitute the total heat requirements:

(1) Heat required in raising the temperature of sludge from about 60° to 212°F; evaporating water from sludge; increasing the water vapor and air temperature of the gas; and increasing the temperature of dried volatiles to the ignition point.

(2) Heat required to raise the temperature of the exhaust gas to the deodorizing temperature.

(3) Heat required to raise the temperature of the air supply required for burning plus the excess air.

(4) Heat losses due to radiation.

(5) Cooling air losses.

(6) Heat required for other endothermic reactions taking place.

The heat content of the organic sludge solids serves to raise the end products of combustion along with the moisture content of the filter cake. The sludge solids draw sufficient heat from the surroundings to reach kindling temperature before combustion can start. When the heat released is sufficient to replace the amount withdrawn, combustion will be maintained. When the quantity of heat released is insufficient to maintain combustion temperature at deodorizing level, heat is recovered from the stack gases and reused, or heat is supplied from an outside source.

Unfortunately, the fuel shortage has cast a new shadow on sludge disposal by incineration. The steadily increasing cost and decreasing availability of gas and oil will affect the sludge disposal options of authorities, at least in an economic sense, since sludge in its liquid form contains far too much water to permit economical incineration.

When the sludge solids content can be increased by dewatering to approximately 30%, it is autocombustible; that is, the heat of combustion of the solids in the sludge generates sufficient heat to evaporate excess water, reach combustion temperature, and compensate for heat losses in the flue gas and through the incinerator walls. Generally, sludge cakes below this solids content will require introduction of auxiliary fuel into the combustion chamber to provide sufficient heat to reach the required combustion temperature. Of course, auxiliary fuel requirements will vary with sludge volatile content, presence of incombustible chemicals, heat loss from the incinerator, and excess air usage.

Table 4 presents the data on auxiliary fuel consumption for sludge incineration in seven cities. The weighted average consumption for the seven cities listed in Table 4 is 51.6 gal/ton. If this figure is rounded to 50 gal/ton nationwide, it yields a ballpark figure of 135,000 gal/day, or over 50 million gal/yr of No. 2 oil (or its equivalent) for sludge incineration based on the 1972 sludge incineration data.

Projected figures for the future are expected to increase dramatically. As cities expand their facilities to accommodate secondary and tertiary treatment, greater volumes of sludges with higher moisture levels and greater chemical contents are being produced. Minneapolis-St. Paul, for example, is expanding its treatment facility to process only 33% more sewage by 1976, but it expects to be consuming 330% more fuel, an increase from 4.5 to 14.9 million gal/yr. Ample evidence indicates that improved treatment facilities will soon be producing a sludge that requires approximately 100 gal of fuel to combust 1 ton of sludge, assuming that current techniques are employed.

Clearly, unless there are modifications or improvements in the technology, the trend toward sludge incineration will be at variance with the fuel conservation goals of the Nation. However, many cities have no feasible alternatives to incineration.

TABLE 4: AUXILIARY FUEL CONSUMPTION FOR SLUDGE INCINERATION IN CITIES DURING 1972[1]

City[2]	Sludge Generation (dry tons/day)	% Solids[3]	Daily Fuel Used[4]	Fuel/Ton Dry Solids[4]
Jersey City, N.J.	15	30 (P)	360	28
Providence, R.I.	30	25 (PS)	850	28
Tonawanda, N.Y. (G)	6	25 (P)	100	17
Rochester, N.Y. (B)	17	30 (P)	715	42
Hartford, Conn. (GS)	25	15 (PS)	4,000	160
St. Louis, Mo. (Lemay)(G)	55	30 (P)	1,200	22
St. Paul, Minn.	275	20 (PS)	12,000	45

[1] Figures were obtained by telephone from plant operators.
[2] G = city uses natural gas; fuel figures are Btu equivalents.
GS = city uses oil in winter and natural gas in summer.
B = oil figure includes fuel for boiler in building.
[3] P = primary only.
PS = primary and secondary.
[4] In gallons of No. 2 oil or the equivalent (if natural gas was used, it was converted into an amount of oil that would yield an equivalent amount of heat).

Options for the Future

Perhaps the best potential method of reducing auxiliary fuel is the incineration of solid waste along with sewage sludge. Refuse as an auxiliary fuel has much going for it. It is a universal commodity found in all communities at generation rates proportional to those of sewage sludge. Refuse is a low sulfur fuel with a heat content sufficient to provide the necessary heat energy to combust sludge.

The same circumstances that tend to push communities toward incineration of sludge also work in favor of refuse incineration, so that the two are most often practiced in parallel. It would seem logical if the two disposal operations were performed concurrently, either in a refuse combustor, a sludge incinerator, or some hybrid unit.

Coal offers the advantage of abundant supply and the prospect of serving as an aid to dewatering. However, use of the most readily available high sulfur coal would depend on the ability of air pollution control equipment to remove sulfur oxides from the exhaust gases. Fortunately, well designed and maintained wet scrubbers used to control particulate emissions have demonstrated the capability to capture over 90% of the sulfur dioxide in flue gases.

Paradoxically, heat treatment of sludge, which requires heat energy, can conserve fuel. Heat-treated sludge dewaters to a high solids content. Excess heat created when this high solids cake is incinerated can be used to provide the heat energy required for the heat treatment. This scheme is not without its drawbacks, because process complexity is increased and substances dissolved during the heat treatment increase the biological load on the treatment plant.

On the horizon are new dewatering processes that may be capable of producing a cake that is autocombustible. Mechanical presses and pressure filters capable of delivering cakes in excess of 50% solids are in various stages of development. Other alternatives would appear to be an improved wet oxidation process and the extensive use of countercurrent heat exchangers in incinerators. Also, combustion techniques such as starved-air incineration may yield a more fuel-efficient operation.

Current Measures to Reduce Auxiliary Fuel Requirement

There are a number of methods that could be employed to reduce fuel consumption: (1) Thermal losses that are due to temperature cycling could be reduced. Thermal losses from shutdown and restart of a multiple-hearth incinerator can account for as much as 15% of the total heat loss and as much as 80% of the auxiliary fuel demand. Therefore, the largest and easiest reduction in consumption can be obtained by eliminating much of the startup-shutdown procedure by adopting multiple-shift, 24-hr per day operation wherever possible.

Since many cities have oversized incinerators that only operate a few days a week, storage capacity for the raw or thickened sludge is required. Cities that have recently converted from digestion to incineration could utilize abandoned digesters for liquid sludge storage. Incinerators could be allowed to cool down completely between burnings. Then, when the storage capacity is exhausted, the dewatering equipment and incinerator(s) could be operated continuously until the inventory of sludge is consumed.

(2) The solids content yield of the dewatering equipment could be increased. As stated earlier, sludge is autothermic at approximately 30% solids. However, above 30% solids there is an incremental reduction in auxiliary fuel demand for every incremental increase in solids content of the feed. Fine tuning the operating conditions of the dewatering equipment can increase the solids content of the sludge cake.

Vacuum filters, the most common dewatering device, will yield a drier cake through more careful control of the types and quantities of conditioning agents. When the conditioning agents are inorganic chemicals such as lime or ferric chloride, however, auxiliary fuel reduction does not follow a one-to-one correspondence with any reduction in moisture content, since the chemical addition decreases the volatile content of the sludge cake.

The vacuum filter operating conditions can be adjusted to the benefit of fuel conservation. A drier cake is achieved if the rotational speed is reduced or if the liquid level in the sludge pan is lowered. Centrifuges will yield a drier cake if throughput rates are reduced.

Before feeding the vacuum filter, a thickening step would be useful. Filter yield is approximately proportional to the solids content of the input sludge. If the sludge is thickened, it will be possible to slow the filter down and achieve the same yield, but with a drier cake.

(3) Combustion of skimmings and/or waste crankcase oil could be considered. Most cities no longer incinerate plant skimmings because of the difficulties in handling this material. Nevertheless, this material is of high fuel value because of its grease content and addition to the sludge cake should be considered.

Sewage treatment plants may be able to purchase waste oil that has been collected from automotive service stations. Since the imposition of a 6¢/gal tax on re-refined lubricating oil in 1965, the market for this commodity has diminished substantially. However, there is currently a 25% shortfall of lubricating oil crudes that could reduce the availability of this fuel. Used lubricating oil contains high levels of lead and other metals so efficient wet scrubbers will be necessary to reduce particulate emissions to acceptable levels.

(4) Digester methane could be used as fuel. Many plants that have converted to incineration retained their digesters and have continued to digest sludges. In many cases, the digester gas is simply discarded or flared. As supplemental fuel becomes increasingly expensive and less available, greater efforts should be made to constructively utilize this wasted fuel.

If it is not practical to utilize the digester gas as an incinerator fuel, consideration should be given to eliminating the digestion operation. Although digestion reduces sludge mass and may even aid dewatering, it reduces the volatile content of the sludge to an extent that the heating value suffers. Digested sludge must be brought to a much higher solids content than raw sludge for autothermic incineration.

PROCESS VARIABLES

Excess Air

Because of the normal variations in the organic characteristics of the sludge and the feed rate, excess air is added to the combustion chamber. The excess air also increases the opportunity of contact between fuel and oxygen which is necessary if combustion is to proceed. To insure complete thermal oxidation, it has been necessary to maintain 50 to 100% excess air over the stoichiometric amount of air required in the combustion zone. This much excess air is undesirable in that it quenches the reaction temperature by acquiring 12 to 24% of the input Btu's to heat the excess air.

If excess air is not supplied for this reason, it may be difficult to maintain the minimum deodorizing temperature. Therefore, a closely controlled minimum excess air flow is desirable for maximum thermal economy. The amount of excess air required varies with the type of burning equipment, the nature of the sludge to be burned, and the disposition of the stack gases. The impact of use of excess air on the cost of fuel in sludge incineration is shown in Figures 8.3 and 8.4.

FIGURE 8.3: THE IMPACT OF EXCESS AIR ON THE COST OF NATURAL GAS IN SLUDGE INCINERATION

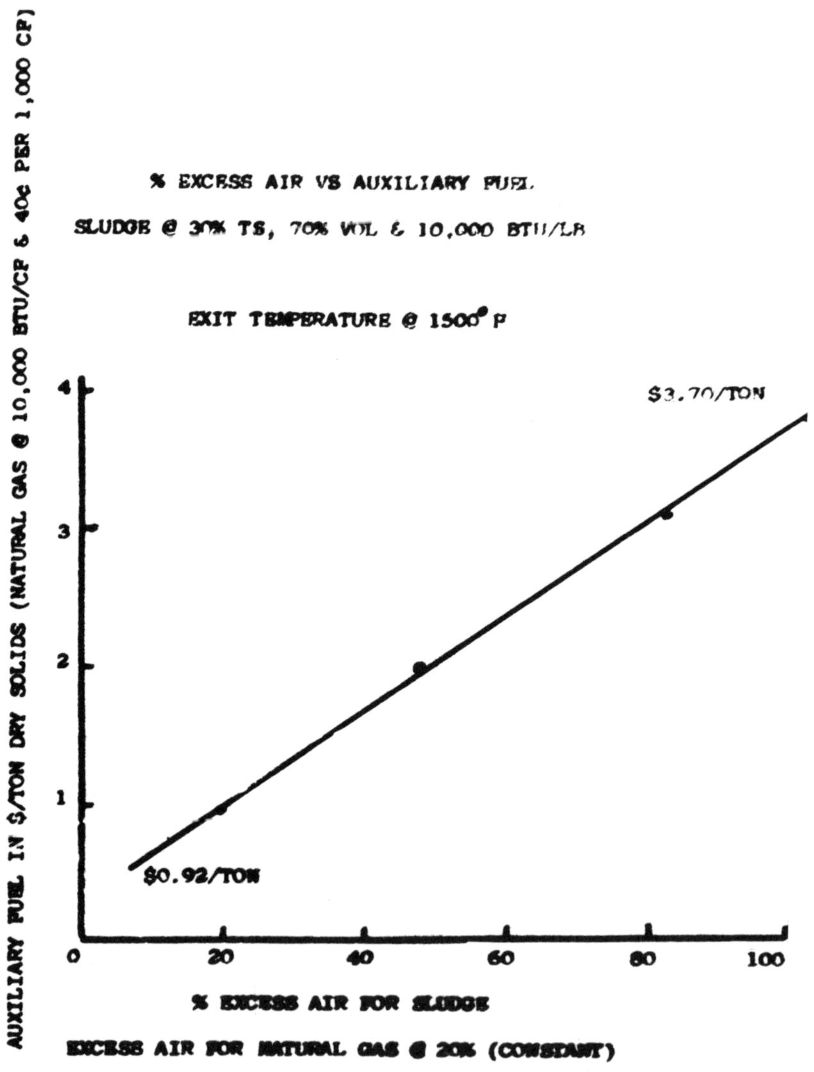

Source: PB 197,888 1970

FIGURE 8.4: THE IMPACT OF EXCESS AIR ON THE COST OF NO. 2 OIL IN SLUDGE INCINERATION

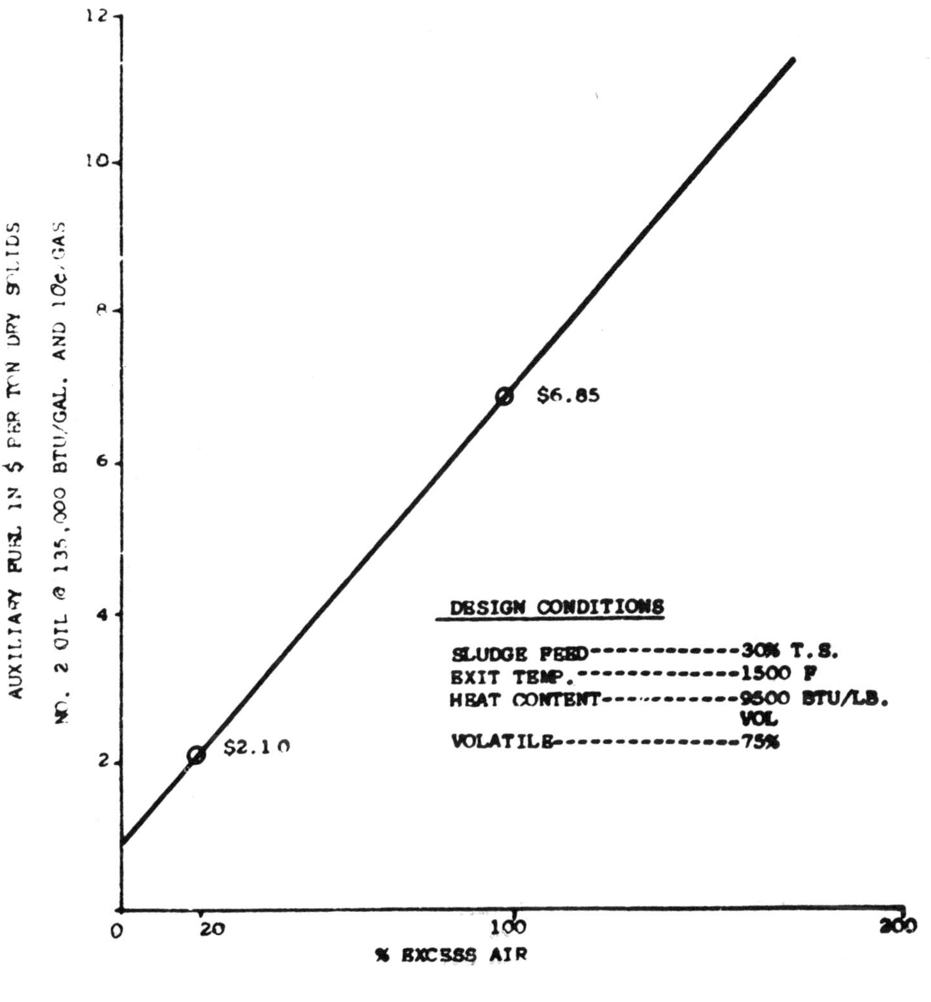

Source: PB 197,888 1970

When the amount of excess air is inadequate, only partial combustion of the carbon occurs, resulting in the formation of carbon monoxide, soot and odorous hydrocarbons in the stack gases. Further, the heat recovered from the partial burning of the carbon is substantially reduced as the heat value of carbon monoxide is only 4,400 Btu/lb.

Preheating and Heat Recovery

Preheating of air is an important step in improving the thermal economy. Air preheat affords an increase in capacity for a given size reactor since the combustion gas volume is used most effectively and since this eliminates the otherwise necessary quantity of auxiliary fuel. The marked effect of preheating air on the cost of auxiliary fuel for various solids concentrations to sustain combustion is shown in Figure 8.5.

FIGURE 8.5: THE EFFECT OF PREHEATING AIR ON FUEL COSTS

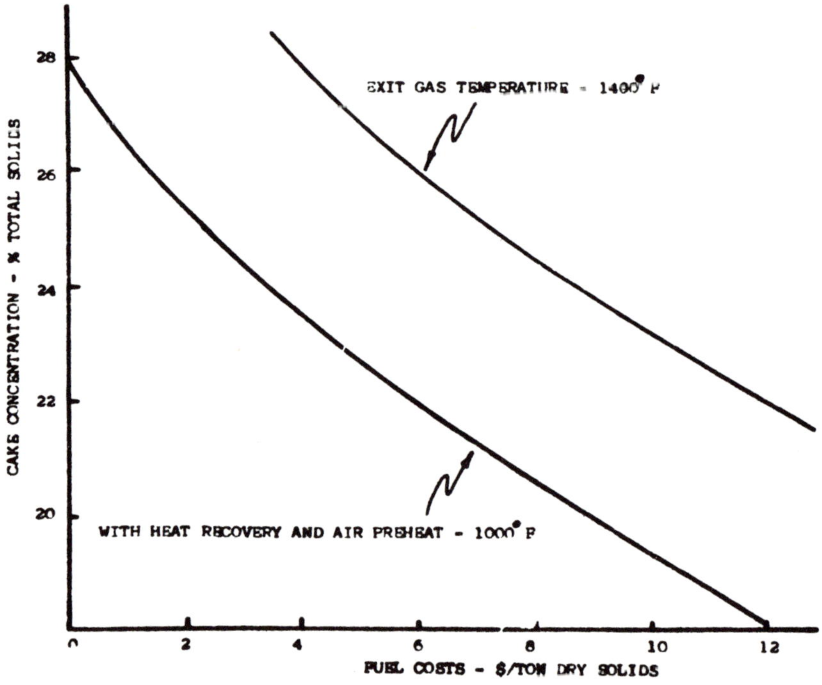

Source: PB 197,888 1970

Preheating of air can be avoided in exceptional circumstances where the following conditions are satisfied:

(1) The excess combustion air volume is maintained at the minimum required to insure combustion.
(2) The grit and inert chemical agents are eliminated.
(3) The moisture content is reduced to a point not often attain-

able by vacuum filtration.

(4) The volatile content of the total solids exceeds 70%.

Heat is recovered from stack gases and the advantage of recovering heat is shown in Figure 8.5. It should be noted that the preheat exchanger represents a significant capital cost and it is to be recommended only after a complete economic evaluation of the process.

Solids and Free Moisture

Most of the sludges to be disposed of by incineration will not support autogenous combustion because of an excessive water content. Thus, auxiliary fuel becomes a prime factor in process evaluation. When the sludge feed is drier, a smaller sized combustion unit is needed and the burning is more efficient. The impact of free moisture on the cost of auxiliary fuel required to sustain combustion for systems with and without heat recovery is shown in Figures 8.6 and 8.7. The importance of obtaining a solids concentration greater than 30% can be illustrated with Figure 8.8. For example, at 25% total solids there is only enough heat available to raise the combustion products and moisture to 900°F and this temperature is far below the accepted 1350° to 1400°F necessary for deodorizing the stack gases of a conventional combustion unit.

FIGURE 8.6: THE EFFECT OF MOISTURE CONTENT ON THE COST OF SLUDGE COMBUSTION

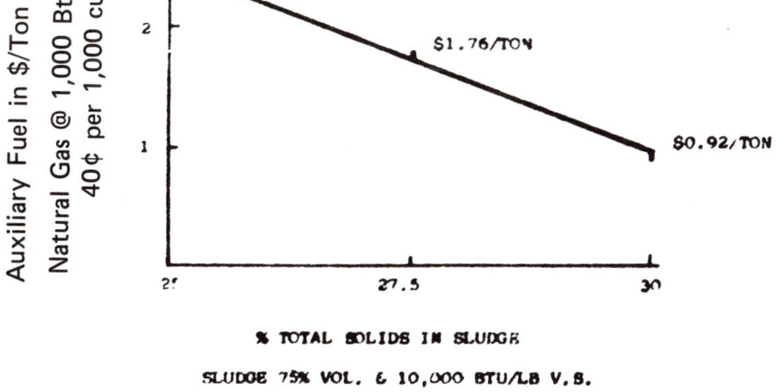

Source: PB 197,888 1970

FIGURE 8.7: EFFECT OF MOISTURE CONTENT ON THE COST OF SLUDGE COMBUSTION SYSTEMS WITH AND WITHOUT HEAT RECOVERY

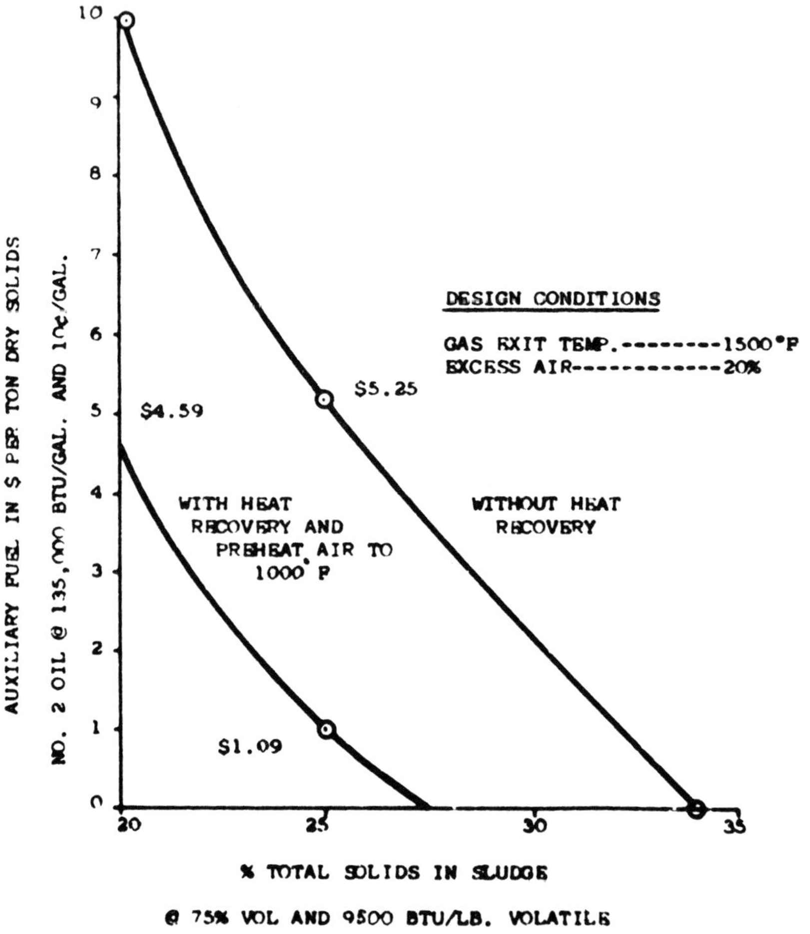

Source: PB 197,888 1970

FIGURE 8.8: EQUILIBRIUM CURVES RELATING COMBUSTION TEMPERATURE TO CAKE CONCENTRATION

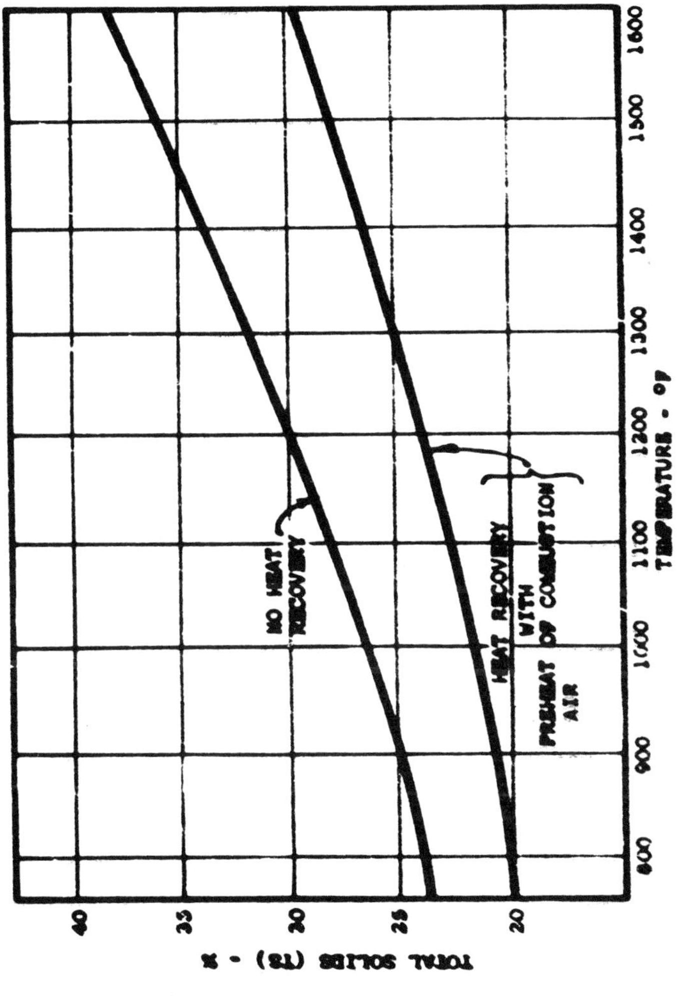

Source: PB 197,888 1970

SLUDGE INCINERATION SYSTEMS

System Components and Makeup

Sludge incineration systems include the following components, in general:

(1) Sludge thickener
(2) A disintegrating or macerating system
(3) Polymer handling and feeding system or other pretreatment schemes
(4) Centrifuge or vacuum filter or any mechanical dewatering system
(5) Incinerator feed system
(6) Air pollution control devices
(7) Ash handling facilities
(8) Complete set of automatic controls such as fail-safe devices, stack temperature regulator and interlocks to permit positive control of excess air.

Incineration processes involve two steps: (1) drying, and (2) combustion. In addition to fuel and air, time, temperature and (for a fluidized bed incinerator) turbulence are necessary for a complete reaction. The drying step is different from preliminary dewatering. The drying is achieved by mechanical means and precedes the incineration process. Sludges having a solids content of 25% and more are delivered to the most common types of incinerators. Because of the high moisture content, the heat required to evaporate the water nearly balances the heat available from combustion of the dry solids.

Drying and combustion is done in separate pieces of equipment or successively in the same unit. Manufacturers have developed widely varying types of sludge drying and combustion equipment. But the principal variation between manufacturers is in the requirement for heating excess air and the efficiency of utilizing the waste gases. The principal types of sludge incineration systems are as follows:

(1) Multiple hearth furnace
(2) Fluidized bed
(3) Flash drying with incineration
(4) Wet oxidation (Zimpro Process)
(5) Atomized suspension technique

Multiple Hearth Furnaces

The most widely used type of incineration system is the multiple hearth furnace. The multiple hearth type of incineration is very popular in large cities where alternate final sludge disposal techniques are inconvenient or too expensive. There are about 120 of these units installed. The types of solids incinerated are very varied.

Multiple hearth units are popular because they are simple, durable and have the flexibility of burning a wide variety of materials even with fluctuations in the feed rate.

A cross section of a typical multiple hearth incinerator is shown in Figure 8.9 and a typical flow diagram for a plant incorporating such a system is shown in Figure 8.10. Multiple hearth units are available in sizes to handle from 5 to 1,250 tons/24 hr. These units are designed with varying diameters from 6'0" to 22'3", and a varying number of hearths, usually between 4 and 12. Multiple units are often used as it allows flexibility of operation. The units are capable of burning grit, screenings, grease and sludge.

FIGURE 8.9: TYPICAL SECTION OF MULTIPLE HEARTH INCINERATOR

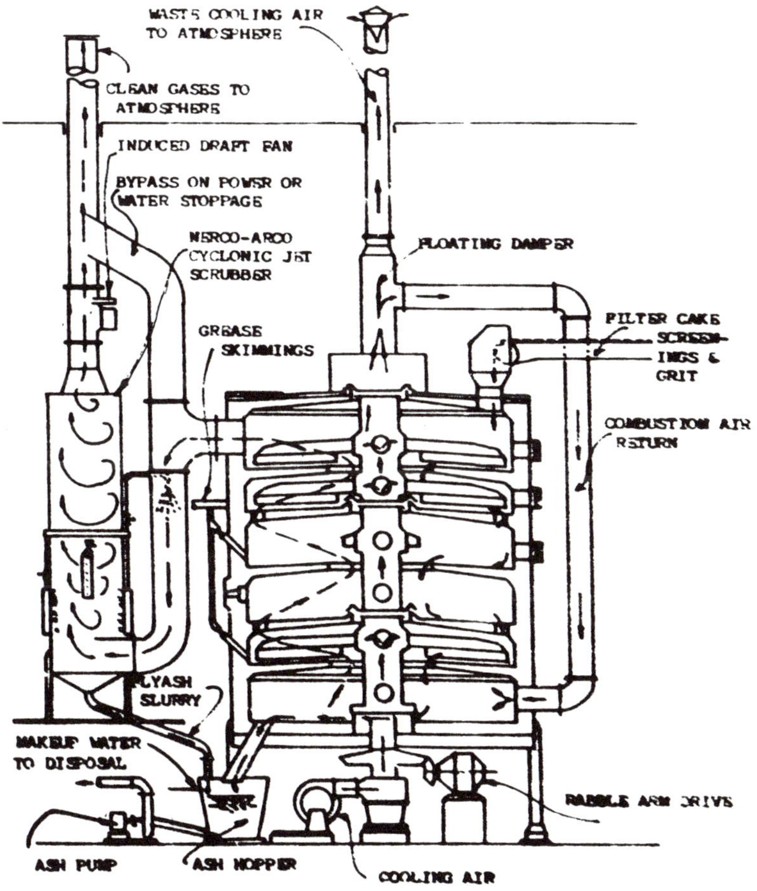

Source: PB 197,888 1970

FIGURE 8.10: FLOW SHEET OF A TYPICAL PLANT WITH MULTIPLE HEARTH INCINERATOR

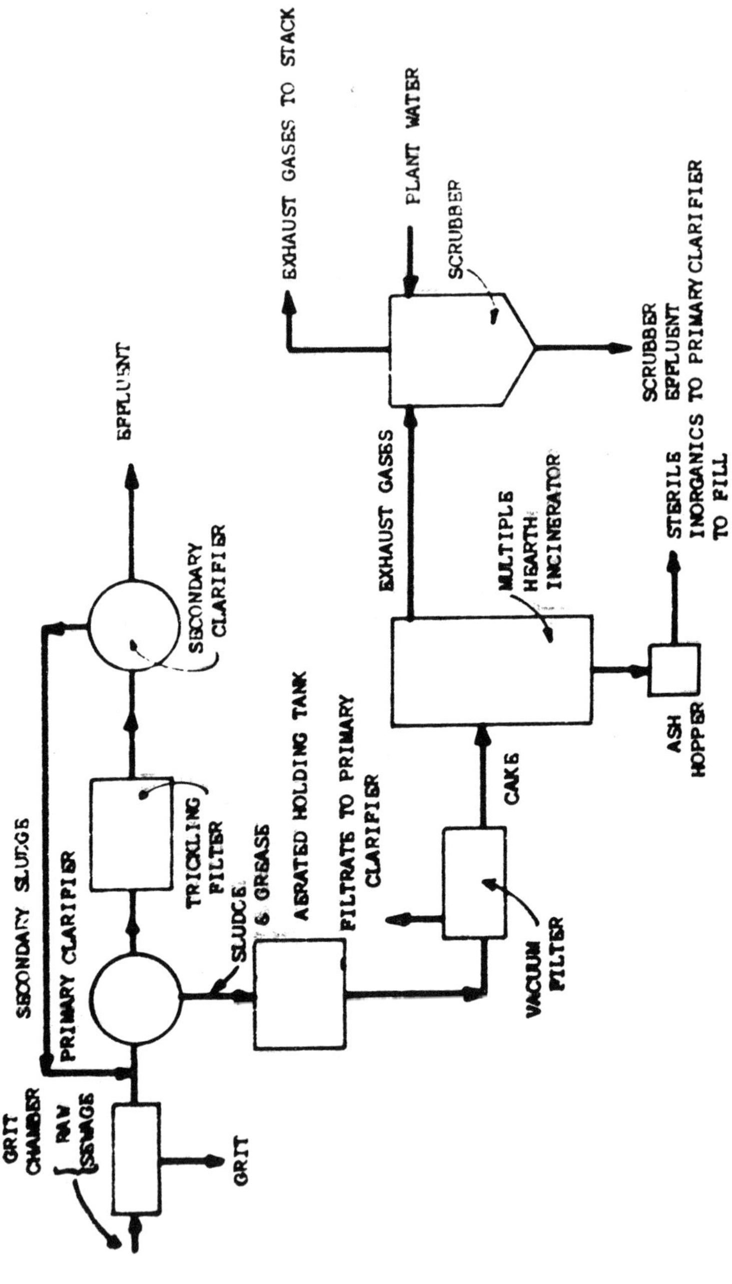

Source: PB 197,888 1970

The design and operation of multiple hearth units are made simple so that they cost less compared to the other types of incinerators. The multiple hearth furnace consists of a circular steel shell surrounding a number of solid refractory hearths (see Figure 8.9) and a central rotating shaft to which rabble arms are attached.

Each hearth has openings that allow the sludge to be dropped to the next lower hearth. The central shaft and rabble arms are cooled by air supplied in regulated quantity and pressure from a blower discharging air into a housing at the bottom of the shaft. Rabbling is very important to combustion because it breaks up large sludge particles, thereby exposing more surface area to the hot furnace gases that induce rapid and complete combustion.

Partially dewatered sludge is continuously fed to the upper hearths which form a drying and cooling zone. In the drying zone, vaporization of some free moisture and cooling of exhaust gases occur by transfer of heat from the hot gases to the sludge. Intermediate hearths form a high temperature burning zone where all volatile gases and solids are burned. The lowest hearth of the combustion zone is the place where most of the total fixed carbon is burned.

The bottom hearth of the furnace functions as a cooling and air preheating zone where ash is cooled by giving up heat to the shaft cooling air which is returned to the furnace in this zone. The temperatures range from 600°F at the bottom, 1600° to 1800°F in the middle hearths, and 1000°F on the top hearths. The waste gases from combustion are heated to deodorizing temperature so as to guard against odor nuisance. Exhaust gases leaving the incinerator at the top are scrubbed in a wet scrubber to remove flyash.

Auxiliary fuel is invariably used to heat the waste gases before venting to the atmosphere and the operating cost becomes expensive due to the fuel cost. A heat recovery device may be suggested to improve the economy of high temperature deodorization but this requires expensive deodorization and combustion air preheating equipment.

Incineration of sludge is generally considered to be more expensive than other sludge disposal processes. However, with the increase in the population served, the economics may be favorable especially for large size communities.

There is a wide variation in operating costs found in the literature and this variation could be attributed to the variation in supplemental fuel requirements. A wide difference between the fuel required for raw sludge incineration and digested sludge incineration was reported and the additional heat requirements for raw sludge and digested sludge were found to be 1.75 and 17.9%, respectively.

Multiple hearth furnaces may also be operated as sludge dryers. The flow of sludge and the rabbling action in multiple hearth furnaces are identical to incineration procedures. Modifications to the basic furnace design include fuel burners at the top and bottom hearths plus down-drafting of the gases. As the solids move downward through the furnace, the gases become cooler and the solids become drier. At the point of exit from the furnace, the gas temperature is about 325°F and the solids temperature about 100°F.

Incineration of sewage sludge in multiple hearth units is progressively increasing and, therefore, some improvements in the design and operation would be desirable even though the units operate quite satisfactorily. Recovery and reuse of heat offers one important potential way of improving the economy of incineration. Recovered heat could be used to condition the sludge to be incinerated. Other areas of improvements could be the development of additional instrumentation to control the combustion process and the development of some beneficial uses for ash, such as the use of ash as an aid to sludge conditioning.

Fluidized Bed Furnaces

Fluidized bed technology, developed for catalyst recovery in oil refining by the Standard Oil Development Company, has been applied to metallurgical roasting, lime mud reburning, spent sulfite liquor combustion, the incineration of municipal and industrial sludges and a host of other industrial applications.

A typical section of the fluid bed reactor used for combustion of sewage sludges is shown in Figure 8.11. The bed material is composed of graded silica sand. When particles are suspended in an upward-moving stream of gases, the mixture of particles and gases behaves much like a fluid. Mixing is an important factor in combustion. The air is supplied as near the surface of the fuel as practical and thoroughly mixed with the combustible matter in order that the combustion may be completed in a short time and the combustion space used more effectively.

Sufficient air is used to keep the sand in suspension but not to carry it out of the reactor. The intense and violent mixing of the solids and gases results in uniform conditions of temperature, composition, and particle size distribution throughout the bed. Heat transfer between the gases and the solids is extremely rapid because of the large surface area available.

The heat required for raising the sludge to the kindling point must come from the combustion zone. While standard combustion units rely on the heat transfer from the hot gases which contain only 16 Btu/cu ft, the expanded bed of the fluid bed reactor has 16,000 Btu/cu ft. Because of the enormous reservoir of heat in the bed and a rapid distribution of fuel and sludge throughout the bed, optimum contact between fuel and oxygen and rapid transfer of heat is insured. The sand bed retains the organic particles until they are reduced to mineral ash and the violent motion of the bed comminutes the ash material, preventing the buildup of clinkers. The resulting fine ash is constantly stripped from the bed by the up-flowing gases.

FIGURE 8.11: TYPICAL SECTION OF A FLUID BED REACTOR (DORR-OLIVER, INC.)

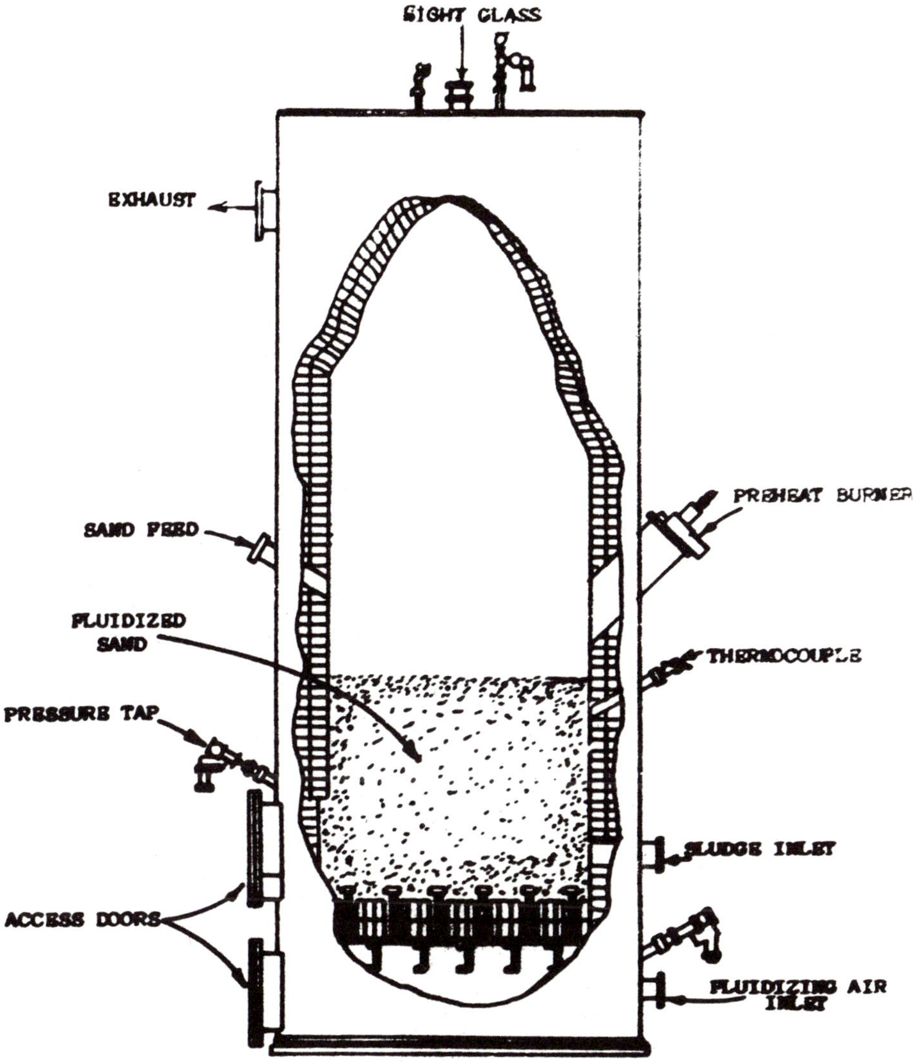

Source: PB 197,888 1970

The major advantages of using the fluid bed reactor include:

(1) Ideal mixing of the sludge and the combustion air is achieved.
(2) Drying and combustion take place concurrently within the bed and, therefore, there is no air pollution problem.
(3) The reactor has no moving parts.
(4) There are no liquid heat exchange surfaces to scale and the operating pressure is as low as 2 psig.
(5) The unit can be operated 4 to 8 hours a day with little reheating when restarting because of the fact that the sand bed serves as a heat reservoir.
(6) Need for a mechanical system for ash removal is eliminated because ash removal from the reactor is accomplished by the upflowing combustion gases.

The flow diagram of the Dorr-Oliver Fluo Solids disposal system is shown in Figure 8.12. The major process steps involved are listed below:

(1) solids preparation
(2) solids dewatering
(3) solids combustion
(4) stack gas treatment.

The solids preparation starts with the degritting operation. Degritting is an important step in order to increase the heat value of the sludge and to protect the equipment from wear and tear. In the conventional design of grit chambers practiced in the United States, usually about 95% of the 48 to 65 mesh particles are removed. To achieve higher degree removal of about 95% of the 200+ mesh particles, hydrocyclones are installed and they represent only a fraction of the installed cost of a gravity unit. Sludge thickening is the next step for solids preparation as it equalizes the sludge flow and increases the solids concentration.

The second step is sludge dewatering and is preceded by a solids disintegrator or comminutor. Sludge dewatering is usually achieved by either a centrifuge or vacuum filter. Dewatering is a very important step as it improves the economics of combustion by reducing the amount of water fed to the reactor.

In the combustion step, the dewatered sludge solids are pumped into the reactor operating at a pressure of about 2 psi and a temperature of 1400° to 1500°F. Sludge has to be fed only when the bed temperature has been raised to 1400°F and this is necessary in order to insure odor control. Because of the high temperature, the sludge quickly dries and burns and, thus, helps maintain the bed temperature. The solids are reduced to inert ash and removed from the fluidized bed by the upward flowing combustion gases.

Exhaust gases are scrubbed usually in wet scrubbing equipment using the wastewater treatment plant effluent as the scrubbing medium. Ash solids are separated

FIGURE 8.12: FLOW DIAGRAM OF DORR-OLIVER'S FS DISPOSAL SYSTEM

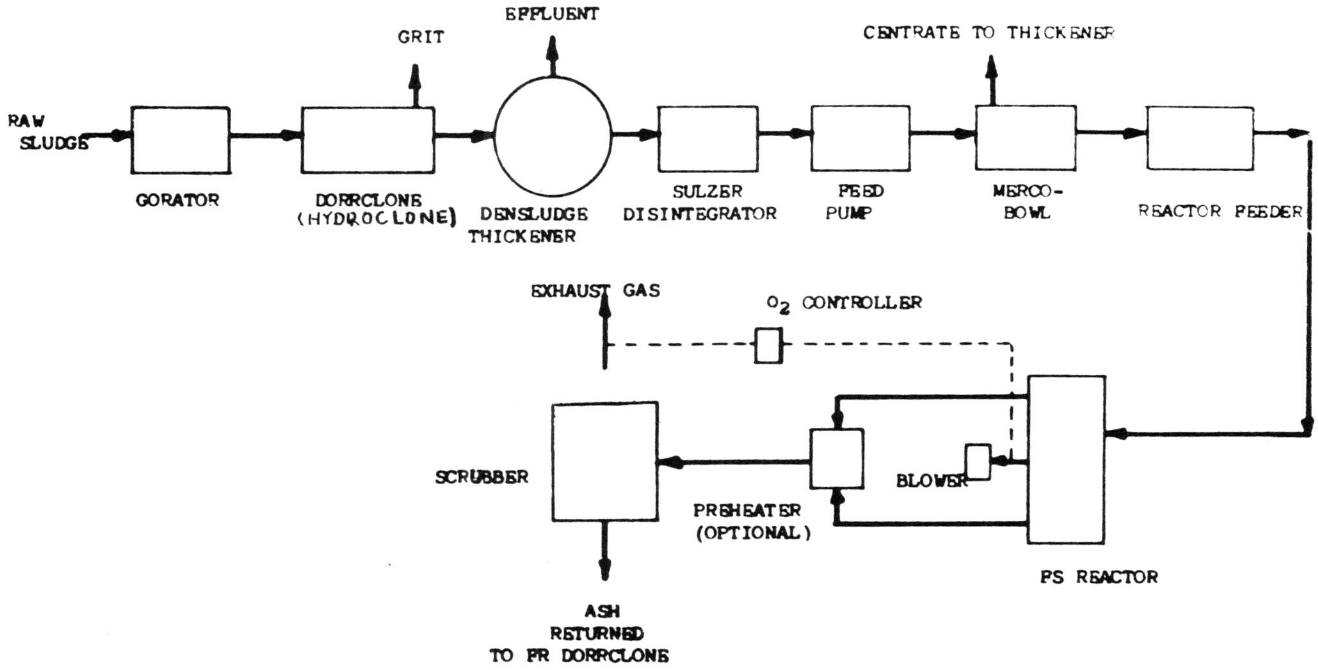

Source: PB 197,888 1970

from the liquid in a hydrocyclone and the liquid returned to the raw waste stream.

Following laboratory test work, a complete pilot plant was installed at the New Rochelle, New York, sewage treatment plant. The initial pilot plant installation consisted of a thickener, a sludge storage tank, a sludge transfer pump, a progressing cavity pump to feed the solids to concentration unit, a screening-type centrifuge, a fluid-bed reactor and a wet impingement scrubber. The recoveries in the screening centrifuge were too low and, therefore, it was replaced by a solid-bowl centrifuge which produced 30 to 40% solids and a recovery of 85 to 95% of the feed solids. The pilot plant tests showed that 10 to 15% excess air was adequate for complete combustion of the carbon and hydrogen.

It was also found that smoke and odor conditions were eliminated above 1100° to 1150°F, compared with the required 1350° to 1400°F for deodorization. However, the combustion capacity of the unit was reduced because of the reduced sludge combustion rate. Radiation losses were found to be less than 4% of the input Btu's when the operating temperature was 1600° to 1700°F.

The first commercial combustion system employing the fluid bed technique was installed in the city of Lynnwood, Washington. Raw primary sewage sludge was combusted in this unit following gravity thickening and centrifuge dewatering. The underflow from the thickener varied considerably and the average concentration was in the range of 10 to 12% total solids.

The scum removed from the primary clarifier is pumped to the thickener along with the settled solids. The floating material removed from the thickener is concentrated by removing the subnatant liquid and, then, by pumping the concentrated scum into the thickener sludge blanket via sludge withdrawal pipe. Pneumatic conveying of the low volatile sludge was found to be impractical and this system was replaced by a stainless steel lift conveyor and retractable extrusion screw.

The fluid bed reactor was designed to receive 220 lb dry solids per hour at 75% volatile sludge at about 35% solids. The reactor has been operated with 20% excess air or about 360 scfm at a sludge feed rate of about 210 lb dried solids per hour. No. 2 oil was used for daily reheating and as auxiliary fuel because the reactor has not been operated continuously. The reheat time and the fuel required for reheating is a function of the duration of shut down and this is shown in Figure 8.13 as observed in the Lynnwood plant.

The feed rate control was automatic and based on the oxygen content of the stack. It has worked very well. The auxiliary fuel system is controlled by the bed temperature, and fuel can be added automatically if the temperature falls below a preset minimum. The system was designed in such a way that it will shut down automatically in case of failure of an item of equipment or instrument.

The scrubber, using about 40 gpm final effluent, was able to cool the exhaust gases to about 160°F in addition to removing the ash. The total operating power

FIGURE 8.13: FUEL REQUIREMENTS FOR PREHEATING AND REHEATING THE LYNWOOD FS REACTOR AFTER VARIOUS SHUT DOWN PERIODS

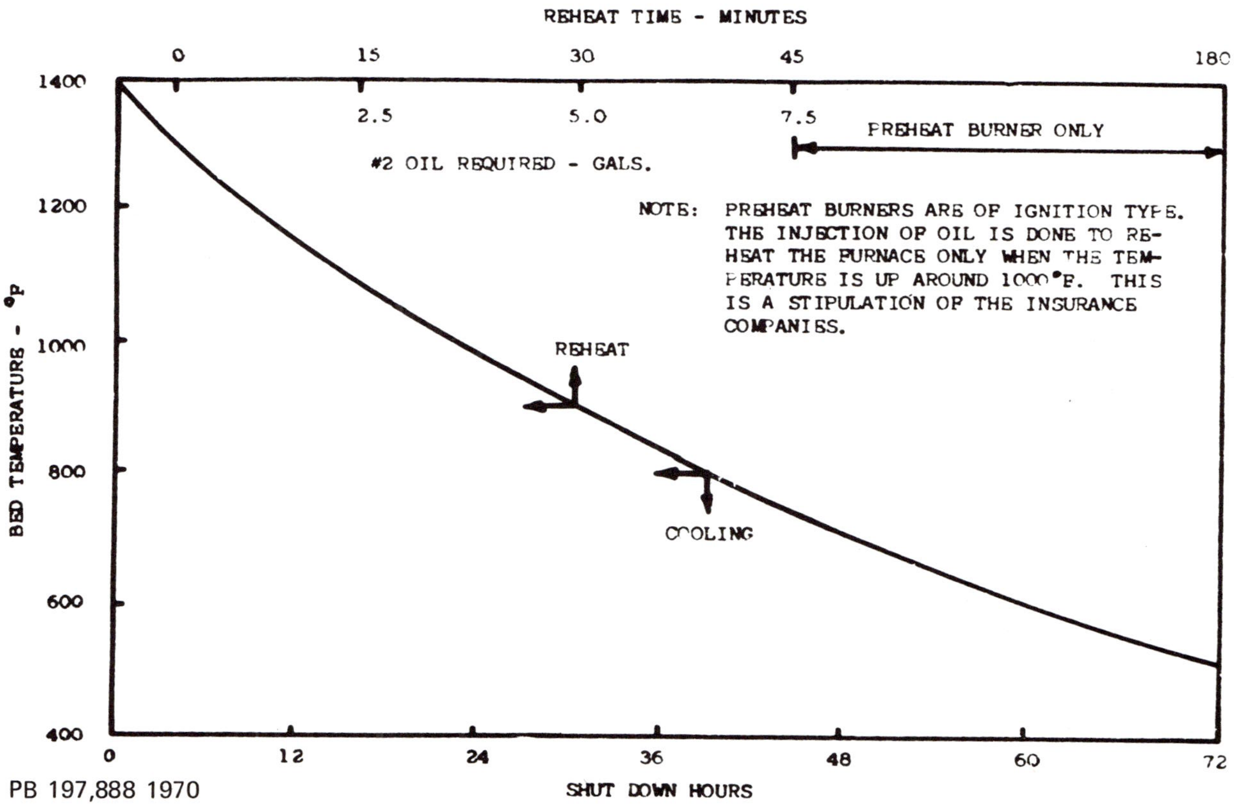

Source: PB 197,888 1970

for the complete disposal system has been estimated at 237 kwh per ton dry solids.

The cost of operation is greatly reduced by the utilization of automatic controls to maintain optimum combustion conditions and, thus, freeing the plant operator for other duties. The results of the pilot plant at New Rochelle and the commercial installation at Lynnwood indicate that high speed centrifugation produces primary sludge concentrations that are economical to burn without heat recovery or the use of a sludge-drying stage.

The economy of the fluidized-bed system is a function of the percentage of the excess air. Therefore, control of the excess air is very necessary and this is achieved by automatic controls.

An air preheater is an optional piece of equipment which will reduce the auxiliary fuel cost. The incoming air for fluidization and combustion is heated to 1000°F by the hot exhaust gases from the reactor. Even though preheating reduces the fuel cost considerably, the capital cost of preheat system is about 15% of the fluidized-bed plant's investment. Further, maintenance cost of the preheating system is considerable compared to other auxiliary equipment.

A study was based on the performance of the plant at Lynwood, Washington, serving a population of 8,000. Cost figures were extrapolated to 22,000 population and it was assumed that volatile matter would not increase above 70%. Power and fuel accounted for 21.7% of the operating cost at the 8,000-population level, and 38.6% at the 22,000-population level. However, based on the total operating cost, including amortization of the capital cost, the figures were more attractive for 22,000 population than 8,000 population.

Feed solids concentration and its volatile content are a factor in the economy of combustion. The higher the solids concentration from vacuum filtration or centrifugation, the lesser the cost of combustion. The effect of percent of volatile solids on the cost of auxiliary fuel for sludge incineration is shown in Figures 8.1 and 8.2. The process steps, degritting, thickening, sludge dewatering, air preheating and close control of excess air, are very essential for optimizing the process efficiency.

Vacuum filtration always requires chemical conditioning whereas centrifugation does not. The operating cost for fuel, power and chemicals for fluidized-bed systems handling raw primary and secondary sludge is $15 to $18 per ton whereas it is $5 per ton for those handling primary raw sludge alone (according to a 1969 report).

The fluidized-bed systems have been operating satisfactorily and they are very competitive with other techniques, especially when the operation is continuous and deodorization is required. When deodorization steps are not required, fluidized-bed systems are more expensive than multiple hearth furnaces.

Flash Drying and/or Incineration

Flash drying is the instantaneous removal of moisture from solids by introducing them into a hot gas stream. This process was first applied to the drying of sewage sludge at the Chicago Sanitary District in 1932 by the Raymond Division of Combustion Engineering, Inc.

The pictorial flow diagram of the C-E Raymond Flash Drying and Incineration System is shown in Figure 8.14. This system is composed of four distinct cycles which can be combined in different arrangements to give the system maximum flexibility to meet specific requirements. The first cycle is the flash drying cycle, consisting of the hot gas duct, cage mill, mixer, uptake duct, cyclone, air lock, dry divider, and vapor fan.

The wet filter cake is blended with some previously dried sludge in a mixer to make it fit for pneumatic conveyance. The blended sludge and the hot gases from the furnace at 1300°F meet ahead of the cage mill and flashing of the water vapor begins. The cage mill mechanically agitates the mixture of sludge and gas and the drying is virtually complete by the time the sludge leaves the cage mill. The sludge, at this stage, is dry at a moisture content of 8 to 10% and the dry sludge is separated from the spent drying gases in a cyclone. The dried sludge can be sent either to fertilizer storage or to the furnace for incineration.

The second cycle is the fuel-burning cycle. Combustion of fuel is essential to provide heat for drying the sludge and the fuel consists of gas, oil, coal or sewage sludge, itself. Primary combustion air, provided by the combustion air fan, is preheated and introduced at a high velocity to promote complete sludge combustion. The sludge ash accumulates in the furnace bottom and is removed periodically by a hydraulic sluicing system to an ash lagoon or other disposal area.

The third cycle is the effluent gas cycle or induced draft cycle consisting of the deodorizing and combustion air preheaters, dust collector, induced draft fan, and stack. Heat recovery is practiced to improve the economy of operation. The effluent gases then pass through a dust collector (dry centrifuge or wet scrubber) and the induced fan discharges the effluent gases through a stack into the atmosphere. The fourth cycle is the fertilizer-handling cycle. Some of the advantages arising out of flexibility in operation are listed below:

(1) Sludge can be dried or incinerated to suit the plant's immediate requirements.

(2) The final moisture content can be automatically very closely controlled since a relatively small amount of sludge is in the system at one time.

(3) The system can be started and shut down in a short period of time and no standby fuel is required when sludge is not being processed.

FIGURE 8.14: PICTORIAL FLOW DIAGRAM OF THE C-E RAYMOND FLASH DRYING AND INCINERATION SYSTEM

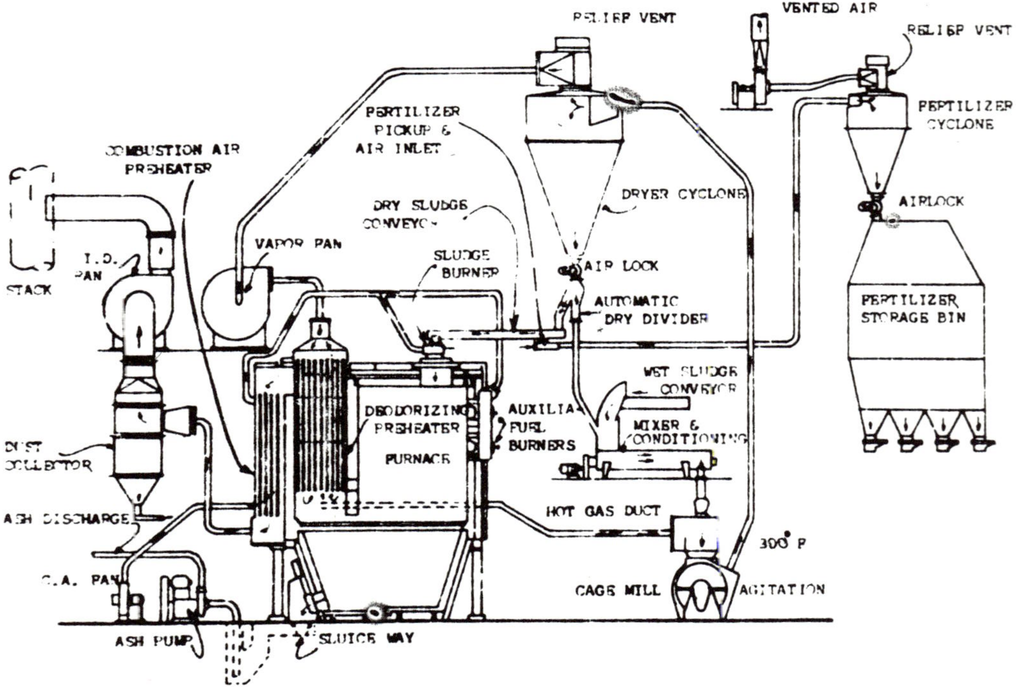

Source: PB 197,888 1970

The lack of fertilizer market for dried sewage has eliminated the major advantage of this system, the flexibility of drying or burning. As an incineration unit, the flash drying system has the major disadvantages of complexity, potential for explosions and potential for air pollution by fine particles. Even though air pollution controls are readily applicable to the flash drying and incineration systems, in comparative situations it is not equal to other furnace designs.

Cyclonic Reactors

Cyclonic reactors are ideally suited for effective and economical sludge disposal in the smaller sewage treatment plants because of their simplicity in installation and flexibility in operation. The mechanism in cyclonic reactors is that high velocity air, preheated with combustion gases, from a burner is introduced tangentially into the cylindrical combustion chamber. Concentrated sludge solids are sprayed radially towards the intensely heated walls of the combustion chamber.

This feed is immediately caught up in the rapid cyclonic flow of hot gases and combustion takes place so rapidly that no material adheres to the walls. The ash residue is carried off in the cyclonic flow and passes out of the reactor. Basically, the performance of the cyclonic reactor depends on (1) the cyclonic flow pattern, (2) the dispersion of the feed, and (3) the temperature of the combustion chamber walls.

Cyclonic reactors have high efficiency operation and this is achieved by the cyclonic action. Cyclonic reactors, manufactured by Dorr-Oliver, are quite compact and occupy a space of about 3.5' x 6' with a height of about 6'. The reactor is brick-lined steel and weighs less than 2 tons. Sargent (Zurn Industries) also produces a cyclonic incinerator.

An atomizing-type oil burner serves as the primary heat source and the hot gases from this burner enter the cyclonic reactor at high temperature and maintain the reactor walls at a high temperature to prevent sludge from sticking while burning on the walls. The sludge is fed into the reactor by a progressive cavity pump and the dispersed sludge particles burn as soon as they hit the wall of the reactor.

These reactors process combined primary plus secondary sludge at a nominal rate of 100 to 130 lb of dry solids per hour or 500 to 650 lb of wet sludge per hour. The detention time for the sludge within the reactor is less than 10 seconds. The cyclonic reactor oxidizes the sludge, producing inert ash, water and carbon dioxide. The trace amounts of sulfur in the sludge and in the fuel oil are oxidized to sulfur dioxide, but this amount is negligible after efficient scrubbing. The temperature is kept above 1400°F so that the organic matter is burned above the odor producing level.

Wet Oxidation

The wet oxidation process is based on the discovery that any substance capable of burning can be oxidized in the presence of liquid water at temperatures between 250° and 700°F. The process is uniquely suited to the treatment of difficult-to-dewater waste liquors and sludges where the solids are but a few percent of the water streams. In general, given the proper temperature, pressure, reaction time and sufficient compressed air or oxygen, any degree of oxidation desired can be accomplished.

This process has been commercialized and patented as Zimpro process. This process has also been known as wet incineration, wet combustion and wet air oxidation processes. Wet air oxidation does not require preliminary dewatering or drying when compared to the conventional flash combustion. Water can be present up to 99% in this process whereas in conventional combustion it must be reduced to about 75% or auxiliary fuel must be added. Far from quenching the reaction, water is essential to wet air oxidation.

Another significant difference is the flameless oxidation of the organics at low temperatures of 300° to 400°F when compared to 1500° to 2700°F in the conventional combustion processes. Air pollution is controlled because the oxidation takes place in water at low temperatures and no flyash, dust, sulfur dioxide or nitrogen oxides are formed.

In the Zimpro continuous process, the sludge is passed through a grinder which reduces any particles greater than ¼" to about ¼" size. Sludge and air are then pumped into the system and the mixture is passed through heat exchangers and brought to the initiating reaction temperature. As oxidation takes place in the reactor, the temperature increases.

The oxidized products leaving the reactor are cooled in the heat exchangers against the entering cold sludge and air. The gases are separated from the liquid carrying the residual oxidized solids and released through a pressure control valve to a catalytic oxidation unit for complete odor control. Where economic conditions make it attractive, the gases may be expanded in power recovery equipment before being discharged.

The oxidized liquid and remaining suspended solids are released through a level control valve and the solids may be separated by settling and drainage in lagoons or beds, or other methods depending upon project requirements.

For startup, heat is obtained from an outside source, usually a small steam generator. With high degree oxidations and high fuel value sludges, no external heat is needed once the process is started. Whenever the process is not thermally self-sustaining, steam may be injected continuously to sustain the reaction temperature.

This process may be applied to any type of sewage sludge, raw or digested, primary or secondary, plus scum and screenings. No special thickening beyond conventional settling is needed for application of the process. The choice of the degree of organic matter destruction, of COD reduction, depends on economic factors such as availability and cost of land, size of plant, power rates, etc.

This process can provide a wide range of oxidation and products depending upon the requirements of the application. It can be designed for high oxidation to produce a minimum volume of inert ash or for low oxidation to produce a residue containing stabilized organic matter with soil conditioning value.

The chemical oxygen demand (COD) of the waste serves as a valuable parameter for the design of the wet air oxidation process and waste sludges most applicable to the process have COD values between 25 and 100 g/l (25,000 to 100,000 mg/l). Zimpro units are operated either on a continuous or batch operation basis. In either case, the basic principles of oxidation are the same and the oxidation achieved depends on the temperature, pressure, holding time in the reactor, and the solids concentration of the sludge entering the process.

The degree and rate of sludge solids oxidation are significantly influenced by the reactor temperature. With increased temperature, a higher degree of oxidation is possible with shorter reaction times. The relative COD reduction with increase in temperature from 100° to 300°C for a number of different sludges is shown in Figure 8.15.

FIGURE 8.15: COD REDUCTION VS TEMPERATURE

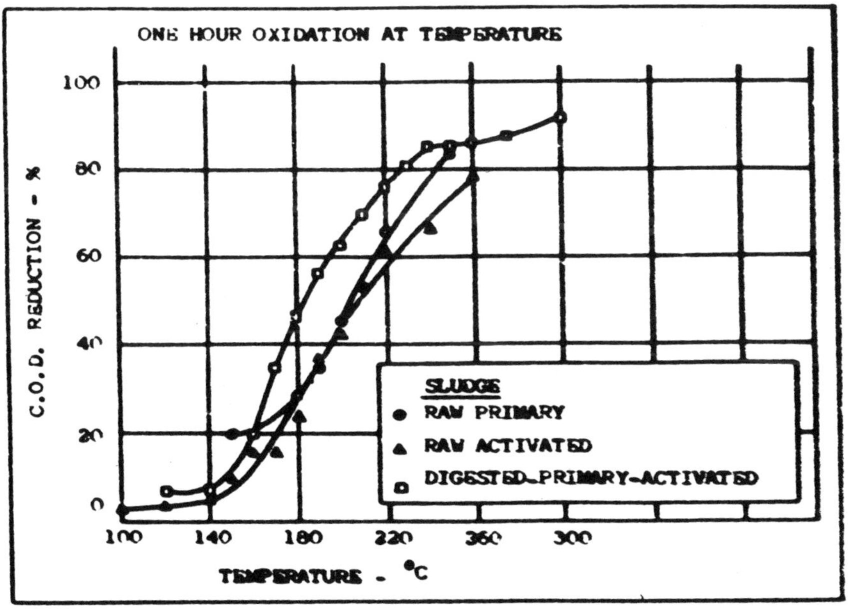

Source: PB 197,888 1970

Oxidation in an aqueous system requires sufficient pressure in the reactor to prevent water vaporization because temperatures are above 212°F. Operating pressures varied from 150 to 3,000 psi, depending on the size of the plant and the degree of oxidation required.

The cost of wet air oxidation can be reduced considerably by increasing the feed sludge solids concentrations to about 6% and future modifications of the heat exchangers and pump capacity may reduce the cost further.

Pre-1969 operating costs of the high pressure Zimpro plant, including labor and maintenance, were about $15.00 per ton of solids processed. The wet air oxidation process has many advantages in sewage sludge disposal and they are as follows:

(1) Sterile end products low in volume and of special value as a soil amendment are produced.

(2) Flyash or dust are not produced as the oxidation takes place in the presence of water. Sulfur dioxide and nitrogen oxides are not formed and odor control is assured by use of gas incineration devices.

(3) Wet air oxidation renders sewage sludge easily dewaterable by filtration or settling without chemical conditioning.

(4) This system permits a clean, sanitary plant without exposure of operating personnel to obnoxious unsterilized sludge solids.

(5) Wet air oxidation has the flexibility of handling any type of sludge and a wide range of oxidation conditions and end products is possible.

(6) Land and building area requirements are minimal and the Zimpro units can be easily integrated into existing facilities with digesters, vacuum filters or incinerators to increase the sludge-handling capacity.

There are disadvantages associated with the wet oxidation process and the major one is the cost of construction and operation. The specific cost depends on the required degree of oxidation which, in turn, depends on factors unique to a local situation such as the size of plant, the availability of land and the cost of power.

Odor problems can develop from the off-gases and from lagooning of the ash-containing effluent. Though air pollution caused by the stack gases can be controlled by catalytic burning at high temperatures, this is an unknown added expense. Another suggested disadvantage of wet combustion systems is the need for high quality supervision and frequent maintenance due to the use of sophisticated equipment and controls.

A major operational disadvantage is the need to recycle wet oxidation liquors back through the wastewater treatment processes. This may represent a considerable organic load on the system and the fine ash could plug air diffusion plates and sludge vacuum filter media. Therefore, the effluent requires further treatment before final disposal. It should also be realized here that the wet oxidation cannot approach the degree of destruction of organic matter as is achieved in a true incineration process.

It is possible that wet air oxidation has the potential of being the best method for ultimate sludge disposal and further research and development of this technique could bring down the capital and operating cost.

Atomized Suspension Technique (AST)

The atomized suspension technique is designed for high tempeature-low pressure thermal processing of wastewater sludges. In this system, sludges are reduced to an innocuous ash and bacteria and odors are destroyed. This system is known under different names such as spray evaporation, and thermosonic reactor system.

Figure 8.16 shows the basic components of the system and the unique features of the atomized sludge incineration process start with a sonic atomizer that produces a mist and fine particle spray at the top of the reactor. The process generally includes the following steps:

(1) Thickening of the feed sludge to 8% and higher.
(2) Grinding the sludge to reduce the particle size to less than 25 microns.
(3) Spraying the sludge into the top of a reactor to form an atomized suspension.
(4) Drying and burning the sludge in the reactor.
(5) Collecting and separating the ash from the hot gases.

The important parameters in the design and performance of atomized suspension incineration include: sludge type, sludge solids concentration, amount of excess air used, pressure in the reactor and sludge particle size. Sewage sludges, in general, are thermally not self-sufficient unless first dewatered in mechanical equipment. It has been estimated that a raw sludge having a heating value of 8,780 Btu/lb of dry solids would have to be thickened to 14% to be thermally self-sufficient.

Particle size distribution is an important factor to be considered to prevent sludge stoppages in lines and in the atomizing nozzles, and to improve the combustion. With the increase in the particle volume, the rates of evaporation and heat transfer in the reactor are increased directly proportional to particle volume.

The operating pressure is kept under 30" of water to prevent leakage from the equipment and to insure no inhibition of evaporation and gasification.

FIGURE 8.16: THERMOSONIC REACTOR SYSTEM FOR TREATMENT AND DISPOSAL OF RAW SLUDGE

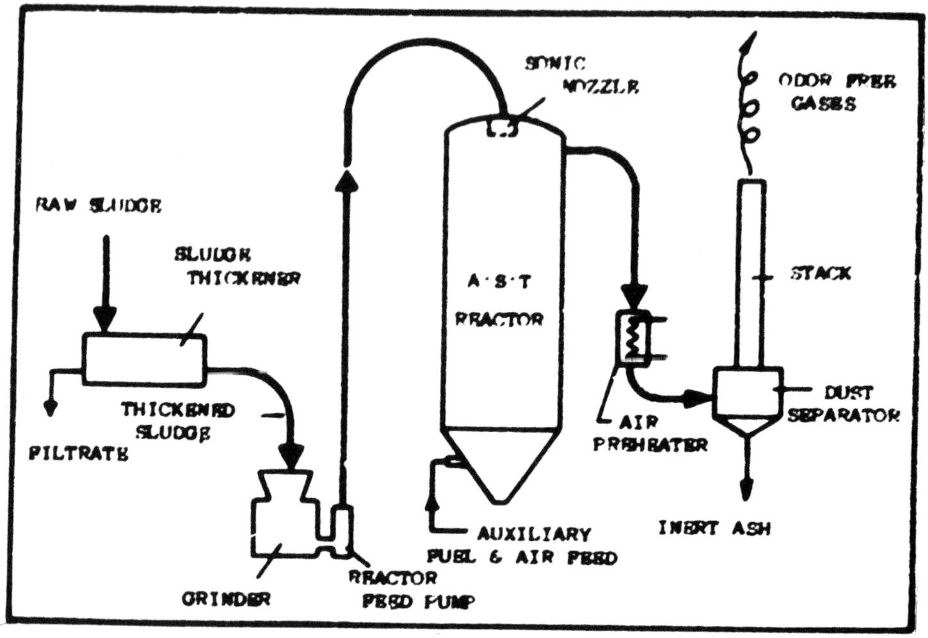

Source: PB 197,888 1970

This system has the following advantages: versatility in sludge handled, small space requirement, rapid conversion of raw sludge to innocuous ash, steam and CO_2, and no nuisance conditions. It has been estimated that the cost will be somewhat higher than conventional incineration processes due to maintenance and the need for supplemental fuel oil or gas.

Capital costs do not also appear to be less than for other incineration techniques. The AST process has an advantage over the Zimpro process in that the operating pressures are much lower. The possibility of incinerating a dilute sludge, thereby eliminating costly dewatering steps, is very attractive.

SPECIFIC EXAMPLES

In the spring of 1971 an EPA task force was charged with the responsibility of determining whether or not sludge incineration would be an acceptable disposal technique, PB 211,323 August 1972 describes its work. Three incinerators were chosen for detailed sampling and analysis. These three were at Barstow, California, South Lake Tahoe, California and Lorton, Virginia. The Barstow incinerator input included only raw sludge from primary treatment; this installation utilized the fluidized bed principle. The South Lake Tahoe unit burned both sludge from the primary process and residue from an activated sludge unit; a multiple hearth incinerator was used. The Lorton, Virginia incinerator was also multiple hearth; it handled primary sludge and grease. A summary of the information which has resulted from the sampling program conducted at the three sites can be given as follows.

Stack Gas Analyses for Particulates, NO_X, SO_2, and Visible Emissions

Analyses of the stack gases indicated that none of the incinerators exceeded an overall dry filterable particulate average concentration of 0.07 grains per standard cubic foot for all test runs inclusive. In the case of the largest incinerator, handling a dry solids feed of approximately 1,225 lb/hr, as a gross average during three test runs, the mass per unit time emission did not exceed an overall average of 2.0 lb/hr.

Oxides of nitrogen emissions were obtained only at South Lake Tahoe and Barstow, California; the higher overall test average emission concentration of the two did not exceed 175 ppm. Sulfur oxides, again, were measured only at the California sites; measured concentrations did not exceed 13 ppm as an average emission during the test period. Visible emissions, at all three test incinerators, were less than 10% opacity.

These test results indicate that well designed and operated sludge incinerators are able to achieve acceptably low emission concentrations for the common pollutants. It should be stated here, however, that the three incinerators tested handled essentially domestic (residential) sewage loads.

Stack Gas Analyses for Major Metals

Measurable concentrations of metals were found in the stack gas material collected from two California sites. Measurable emissions on a mass per unit time basis included 0.06 lb/hr, as a gross average of three test runs, for lead.

Input Sludge Analyses

As would be expected from the stack gas samples, an examination of the input sludge showed measurable quantities of the heavy metals and other important pollutant compounds. The elements cadmium, beryllium, nickel, copper, lead, chromium, vanadium, and mercury are said to be toxic and accumulate in the human body; all, with exception of beryllium, nickel, and vanadium were found in input sludge analyses made at Barstow, South Lake Tahoe, and Lorton. Vanadium was found only at Barstow.

In addition, analyses were made for major pesticides in the input sludge at the three incinerator locations. At Barstow, measurable concentrations of PCB, dieldrin, chlordane, and DDD were found. At Lorton, measurable concentrations of all of the above were uncovered. At South Lake Tahoe, only PCB was analyzed for, and was found.

Analyses of Incinerator Ash

For South Lake Tahoe, Barstow, and Lorton, metal analyses were also made of the ash remaining after incineration of the sewage sludge. The data showed that all of the mercury in the sludge is apparently lost in the stack gases, and that lead compounds are apparently carried off in the fly ash. A comparison of sludge ash composition with a coal fly ash showed the sludge ash to have concentrations of copper, zinc, chromium and lead one or two orders of magnitude higher than coal fly ash. No pesticide or PCB materials were found in the ash or scrubber water.

It was the opinion of the Office of Research and Monitoring solid waste representative on the Task Force that ash resulting from the incineration of sewage sludge would constitute no significant threat to the environment if ultimate disposal of the ash was by acceptable sanitary landfill practice as outlined by *Sanitary Landfill Design and Operation,* a United States Environmental Protection Agency report.

Because of the importance which was given to the measured presence of various metals found in samples taken at South Lake Tahoe, Barstow, and Lorton, sludge samples were taken at additional sewage treatment plants, and analyzed for metals content. The added test sites were at Monterey, California, and at the Middlesex County, Joint Meeting, Passaic Valley, and Bergen County sewage treatment facilities in New Jersey. The New Jersey plants included units with significant industrial inputs.

Again, tests at the additional plants indicated the presence of biologically nondegradable metals. Quantitative results of these tests, however, were available only in raw data form when this report was written and are therefore not included here. Based upon relationships developed during the comprehensive sampling program carried out at South Lake Tahoe, Barstow, and Lorton, it can be expected that quantities of certain heavy metals could be found in the incinerator effluent of the sludges analyzed from Monterey and the New Jersey plants.

The Monterey plant was also tested for stack effluents. Basic results were similar to those obtained from South Lake Tahoe, Barstow, and Lorton.

Conclusions

(1) When operated properly, today's sludge incinerator systems, which have been designed to meet existing air quality standards, have been shown by test to produce acceptable stack emissions of particulate matter, nitrogen oxides, sulfur oxides and odors. This was true for both fluidized bed and multiple

hearth type incinerators. Most sludge incinerators which are in existence today, however, do not incorporate high efficiency particulate matter control devices.

(2) Small, but measurable, quantities of specific toxic metals which are known to accumulate in the human system, were found in the input sludge, stack emissions, scrubber water, and residue of those incinerators which were subjected to comprehensive testing under Task Force supervision. These same metals were also found in each instance where only the sludge alone was analyzed.

(3) Small, but measurable, quantities of specific organic chemical compounds including various pesticides and polychlorinated biphenyls, which are known to accumulate in the human system, were found in all of the sludge samples analyzed. It should be expected that, under conditions of poor combustion, such compounds could be emitted from the stacks of sewage sludge incinerators.

(4) It was impossible for the Task Force to accurately establish the potential for health effects which might be associated with sewage sludge incineration because:

(a) There are insufficient health effects data relating to low atmospheric concentrations of suspected pollutant materials including certain elements such as cadmium, arsenic, beryllium, nickel, copper, lead, chromium, vanadium, and mercury, organic pesticides, and polychlorinated biphenyls.

(b) There are insufficient stack gas sampling and analysis methods sophisticated enough to produce accurate information regarding the quantity, size distribution, and constituent quality related to size, of the particulate matter emitted by sewage sludge incinerators.

Incineration of Oil Refinery Wastes

The material in this chapter is based upon EPA PB 202,570 March 1971 prepared by the American Oil Company.

Among the many waste disposal problems incident to petroleum refinery operations, the American Oil Refinery at Mandan, North Dakota, found the problem of disposing of oily sludge and spent caustic most perplexing.

Oily sludge from refinery operations is characterized by the presence of oil, solids, and water. Its consistency varies from that of a fluid slurry to a greasy semisolid. It is encountered when cleaning oil storage vessels, removed from the waste water oil-water separator and it is collected as a residue from oil-water emulsion breaking operations.

Disposition of this material can include dumping into open earthen sludge pits. In order to conserve land, the volume of sludge is reduced by burning the hydrocarbon that eventually migrated to the surface of the pit. Burning produces a billowing black cloud of smoke. This method of disposing of sludge was most unsatisfactory; it degraded the land and created an obvious source of atmospheric pollution.

Spent caustic results from the use of solutions of caustic soda to treat petroleum products for the removal of acidic impurities such as naphthenic acid, cresols, mercaptans, and hydrogen sulfide. These compounds emit intense disagreeable odors. When an attempt was made to dispose of spent caustic by neutralization and by trickling the neutralized waste into the waste water treating system, it was found that the odor released during neutralization was intolerable and the resulting neutralized waste overtaxed the biooxidation pond, thus degrading the quality of the waste water effluent from the refinery.

In the long term effort to find a solution to these waste disposal problems, various incineration systems have been considered. The oxidation of spent caustic alone would require an economically prohibitive quantity of supplemental fuel and it would be difficult to prevent atmospheric pollution. The incineration of oily sludge appeared more feasible. However, when oily sludge is subjected to high temperatures, the oil evaporates. Should the resulting hydrocarbon vapors fail to ignite, they could create an extremely hazardous explosive mixture with combustion air.

Conditions offered by the fluidized bed incinerator include extremely high heat transfer efficiency, excellent mixing, and stable combustion conditions made the system most promising. Furthermore, a bed of inert fluidized solids acting as a heat sink reduced the hazard created by a mixture of unburned hydrocarbon vapor and air.

Fluidized bed incineration has been successfully applied to the disposal of paper mill sulfite liquors. The process based on research conducted at the Battelle Memorial Institute in Columbus, Ohio, and developed by the Copeland Process Corporation of Oak Brook, Illinois, appeared to be adaptable to the disposal of petroleum refinery generated spent caustic and oily sludge.

DESIGN OF THE UNIT

Under contract with the American Oil Company, the Copeland Process Corp. designed and constructed the fluid bed incinerator. It was designed to operate for 6 hours a day during 9 months of the year disposing of the normal production of sludge and spent caustic. During the 3-month period in which tanks are cleaned, the incinerator would be in operation 24 hours a day.

Samples of typical sludges and spent caustic were analyzed. By pilot plant operations, it was confirmed that the spent caustic and oily sludge from the Mandan Refinery was amenable to fluidized bed incineration. However, the pilot plant exhaust gas line plugged during the test. The plugging was believed to be caused by the presence of iron sulfate. Evidently the iron was present in the sludge as rust from tank bottoms. Based on this data, the system was designed to operate with a controlled reactor exhaust gas temperature to prevent scale in the exhaust duct. It was recommended that the incinerator be operated at a bed temperature of 1300° to 1500°F.

Salient design features incorporated in this unit which differ from the typical application of fluidized bed incineration to paper mill waste sulfite liquors were as follows:

(1) Waste feed injection – The feed to the waste sulfite liquor incinerator is sprayed into the vessel from the top. The spray cools the effluent flue gases as it evaporates. The dried solids fall to the fluidized bed where they are oxidized. The

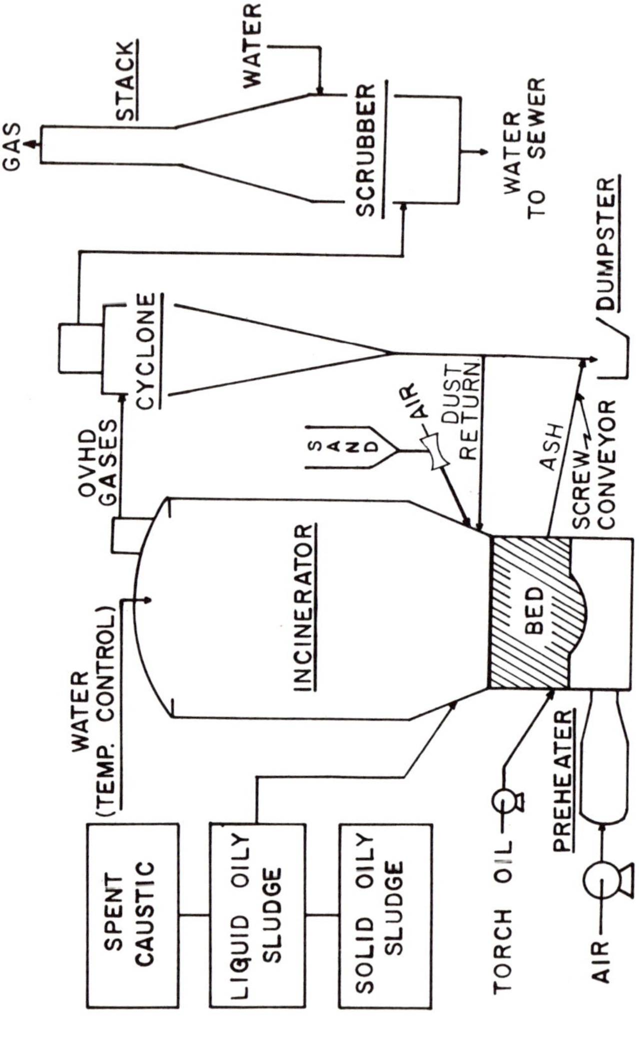

FIGURE 9.1: FLUIDIZED BED INCINERATOR

oily sludge-spent caustic incinerator was designed to inject the mixed feed through a special Copeland burner nozzle just above the fluidized bed of solids. The flue gas temperature was to be controlled by injecting a cooling water spray through the top of the incinerator vessel.

(2) Supplemental fuel for start-up – The Sulfite Liquor incinerators are brought up to operating temperature by preheating combustion air with a natural gas burner. Because of the possibility of creating an explosive mixture of hydrocarbon vapor from the sludge and combustion air, the sludge-caustic incinerator design includes an additional system for injecting Torch Oil (heavy fuel oil) into the bed of fluidized solids a few inches above the bed supporting grid.

When the fluid bed becomes hot enough for ignition, via the natural gas combustion air preheater, the Torch Oil is injected. It is thus possible to maintain a steady high temperature in the bed affording assurance of ideal oxidizing conditions when the sludge is introduced.

(3) The inventory of material comprising the fluidized bed in the unit is normally determined by measuring the differential pressure between a point just above the grid and a point in the free board space above the bed. The Mandan incinerator design included an additional pressure tap so that by measuring the differential pressure across a portion of bed, its density could be determined. With knowledge of the bed density, the height of the fluid bed level above the grid can be determined from the differential pressure across the entire bed.

(4) In order to cope with semisolid sludges from tank cleaning, a steam heated 80 bbl sump was included in the design. Facilities were included to stir the contents of the sump, mix it with fluid sludge from the feed tank, and to deliver the mixture to the feed tank.

Figure 9.1 reflects the Mandan fluid bed incinerator process flow as it was designed.

STARTUP AND INITIAL OPERATIONS

To put the unit into operation, the air compressor is started. The air preheater is fired, and as the system is warmed, sand is introduced, forming a bed of fluidized solids. When sufficient sand has been introduced and the fluid bed is in the range of 900° to 1000°F, torch oil is admitted. The burning of torch oil causes a rapid rise in bed temperature. When the bed temperature is in the range of 1200° to 1300°F, liquid oily sludge is introduced. Provided the oily sludge contains a sufficient quantity of combustible materials, air heating and torch oil injection can be terminated. The temperature of the bed is then controlled by balancing feed rate and temperature-control water. When the operation on oily sludge has been stabilized, spent caustic is introduced.

In order for the operation to be self-sustaining, the oily sludge must contain a sufficient quantity of combustible material. About 29,000 Btu are required per gallon of total feed. The burning of oily sludge with a low heating value results in a lower spent caustic to oily sludge ratio.

As the oily sludge is burned, the solids remain in the bed, while the gaseous products of combustion and water vapor discharge through the gas-cleaning system. Water in the caustic solution is vaporized and the combustible material is oxidized; the solids accumulate in the fluid bed. The sodium oxide formed reacts with the products of combustion to form sodium carbonate and sodium sulfate, both of which remain in the fluid bed.

The proper bed level is maintained by withdrawing ash as it accumulates from the deposition of solids. Ultimately, the original sand bed is completely displaced by the incoming solids and chemicals. The overhead gases pass through a cyclone separator from which recovered fines can be either returned to the bed or discharged along with the excess bed material. From the cyclone separator, the gases are passed through a water scrubber before being admitted to the atmosphere.

DESIGN CHANGES BASED ON OPERATING EXPERIENCE

Wind Box Insulation

The unit was designed to conduct the hot air from the air preheater through a chamber (wind box) and air distribution grid before reaching the bottom of the fluid bed. The startup procedure calls for heating the fluid bed to 900°F via the air preheater before injecting torch oil. During the initial attempt to start up, excessive heat loss from the wind box was observed. When the wind box temperature reached 840°F the fluid bed temperature was only 540°F. The unit was shut down and additional insulation was applied to the interior of the wind box. In subsequent startup operations, satisfactory conditions were observed.

Air Distributor Grid Gussets

The air distributor grid was held in place by welding it to the shell of the vessel completely around its circumference. Additional support was provided by 12 equally spaced 1′ - 1½" x 1′ - 1½" gussets made of one-half inch steel plate. These gussets were greater in number and size than was necessary to hold the grid in place. Furthermore, they transmitted excessive heat from the grid to the shell of the incinerator vessel. When the air preheater was in operation, the exterior skin temperature of the wind box reached 850°F ultimately causing stress cracks in the vessel shell. Copeland redesigned the gussets reducing their number and size. After these revisions were completed, the maximum wind box exterior shell temperature observed while the air preheater was in operation was 240°F.

Feed Preparation

Feed handling facilities as built included a 1,000 bbl feed mix tank, a 405 bbl spent caustic storage tank, and an 80 bbl tank bottoms receiving sump. It was proposed that daily pumping of sludge from emulsion breaking and from the waste water-oil separator would be received directly into the 1,000 bbl feed mix tank. Spent caustic was to be transferred to the feed tank in batches based on the estimated oil content of the final mix.

During early operations, only oily sludge was fed to the incinerator. It was immediately apparent that the oil content (heating value) of the feed was critical. Furthermore, the oil content of sludges was observed to vary widely.

Before attempting to charge spent caustic, temporary piping was installed to permit its introduction as a separate stream. The caustic was moved by the spent caustic transfer pump and the flow rate was determined by tank gauge difference. The charge of caustic was adjusted by a hand operated valve to balance the heat released by sludge burning. With this system, it was difficult to adjust and measure the spent caustic charge but it was possible to dispose of spent caustic in this manner.

An attempt was made to operate as originally planned, by mixing caustic with sludge in the feed tank and charging a single stream to the incinerator. In order to dispose of caustic at a maximum rate, it was necessary to blend to the minimum oil content of total feed. A subsequent pumping of separator sludge that was low in oil content caused the mixed feed to fall below the minimum heating value necessary to sustain combustion in the incinerator. Feed to the unit was discontinued and the feed tank mixer was turned off. Some "free water" settled to the bottom of the feed tank and was drained to the sewer thus increasing the concentration of oil in the feed mix. While the move alleviated the problem created by having a low oil content feed, spent caustic drained to the sewer along with the free water.

From this experience it was concluded that spent caustic should be fed to the incinerator as a separate stream so that the caustic feed rate could be adjusted to compensate for the varying oil content of the sludge. The Mandan incinerator is currently operating with the temporary caustic feed system. Future caustic and sludge incinerator designs should provide systems to feed caustic and sludge separately.

Oily Sludge Injection

Based on fluid bed incinerator operating experience in the paper industry, the unit was designed so that sludge feed would be injected just above the fluid bed. When oily sludge was fed to the unit at this location, the oil in the sludge immediately evaporated and burned in the space above the fluid bed, thus creating a condition similar to afterburning. Heat transfer to the fluid bed was poor. Furthermore, because of the high effluent gas temperature the cooling spray water flow was automatically increased, further cooling the fluid bed. While charging sludge above the fluid bed level, it was necessary to provide supplemental heat with the air preheater or by torch oil injection to maintain the proper fluid bed temperature.

When the sludge feed piping was rerouted to permit the injection of sludge through the torch oil nozzles (into the fluid bed a few inches above the grid), the bed temperature rose, cooling spray water was reduced and it was possible to maintain the proper fluid bed temperature without supplemental heat.

Spent Caustic Injection

In an effort to find the optimum point relative to the fluid bed for injecting spent caustic, four conditions were examined, namely:

(1) All spent caustic was injected above the fluid bed.

(2) All spent caustic was injected into the fluid bed as a separate stream.

(3) Spent caustic was mixed with sludge feed for injection into the bed.

(4) Spent caustic feed was split injecting part above the bed and part into the bed.

The rate at which spent caustic can be charged to the unit is limited by the heat content of the sludge. From the standpoint of maximizing the spent caustic disposal rate, there appeared to be an advantage to injecting the caustic above the fluid bed. Spent caustic was charged at a rate producing 0.042 gal of total feed per 1,000 Btu when injected above the bed. A rate of only 0.039 gal of total feed per 1,000 Btu was attainable while injecting caustic into the bed.

However, when caustic was injected above the bed at a rate in excess of 1 bbl per hour, bed particles adhered to the interior of the flue gas exit duct and the cyclone causing a high pressure drop across the cyclone. Successful operations were sustained by injecting up to 1 bbl per hour above the bed and the remainder of the caustic feed into the bed.

Subsequent investigation revealed that the average chemical oxygen demand of the effluent water from the flue gas scrubber dropped by about 100 ppm when the spent caustic injection was moved from the over-bed nozzle to the in-bed nozzle. This amounts to a reduction of approximately 84 lb of chemical oxygen demand per day in scrubber water going to the refinery waste water treating facilities.

There appeared to be no difference in unit performance between operation with a single mixed feed of oily sludge and spent caustic injection into the bed vs separate caustic and sludge feed streams into the fluid bed. Problems associated with mixing the caustic and sludge were previously discussed. From these observations, it was concluded that the separate injection of spent caustic into the fluid bed produces the preferred performance of the unit.

Cooling Water Spray

The temperature of gases above the fluid bed is automatically controlled by spraying water into the top of the incinerator. Initially the unit was equipped with a one-inch nozzle that produced a spray angle of 15° at 15 psi. This spray nozzle permitted unevaporated water to reach the top of the bed. The fluid bed was thus cooled necessitating the burning of supplemental fuel to maintain proper combustion temperature in the bed.

By reducing the nozzle size from one-half inch to three-eighths inch and increasing the angle to 120°, the most satisfactory results were obtained. The smaller nozzle produced a fog type spray, thus preventing the impingement of unevaporated water against the vessel wall and preventing droplets of water from reaching the fluid bed.

It is possible that the problems associated with the injection of spent caustic solution above the bed might have been solved by injecting the caustic at the proper location through a fogging spray nozzle. Some incentive exists for investigating this matter further.

It is noteworthy that Copeland recommended setting the free board temperature controller at 900°F while permitting the fluid bed to approach 1500°F. In practice it was found that free board cooling by the spray water has a tendency to cool the fluid bed even under the best cooling spray conditions. The Mandan incinerator is operated with the free board temperature controller set to hold a temperature approximately the same as the fluid bed temperature.

Feed Distribution

The sludge and caustic were to be injected through a single nozzle above the fluid bed and torch oil was to be injected through a header with four equally spaced nozzles into the bed. Early experience demonstrated that a single torch oil nozzle was all that was necessary or desired. However, when sludge was fed through the torch oil system into the bed, the charge rate was limited by apparent afterburning when using a single nozzle. The operation was considerably improved by using two of the torch oil nozzles for sludge injection.

When using two injection nozzles off of a single header, as one nozzle begins to become plugged the flow simply increased through the other nozzle. It is obvious to operators when one of the nozzles becomes plugged because when this happens there is an immediate rise in free board temperature, an automatic increase in cooling spray water rate, and a decrease in bed temperature. The distribution of sludge feed into the bed through several injection nozzles is obviously necessary. It would be highly desirable to arrange the feed system to insure the equal distribution of charge through each of the feed nozzles.

Scrubber Water

In the original design of the unit, water from the refinery circulating cooling water system was utilized in the venturi-type gas scrubber. Subsequently a piping change was made permitting the substitution of raw (untreated) river water for cooling water to the scrubber. This change reduced the incinerator operating cost by eliminating the water treating chemicals which were lost via the scrubber effluent. Furthermore the chemical oxygen demand (COD) and dissolved solids (DS) content are greater in circulating cooling water than in raw water thus an equivalent reduction in COD and DS was made in the scrubber effluent water.

The scrubber effluent water contributes dissolved and suspended solids to the refinery waste water treating facilities. The amount of these materials depends on operating conditions and cyclone efficiency. In the event that these materials overtax waste water treating facilities, it is suggested that they can be greatly reduced by circulating the scrubber water.

PROCESS LIMITATIONS

Fusion Temperature of the Fluid Bed Material

The incinerator bed material primarily consists of particulate matter from the sludge plus carbonate and sulfate salts formed by chemical reactions of the spent caustic and combustion gases. In the event the bed temperature is permitted to exceed the fusion temperature of the bed material the particles become sticky. They adhere to the wall of the incinerator vessel and form large agglomerated masses in the bed. Ultimately the bed loses fluidity and it becomes necessary to shut the unit down to remove the large chunks of agglomerated bed material.

The bed material fusion temperature is periodically determined in the laboratory. The fusion temperatures as determined by test, ranges from 1000° to 1700°F. It has been found that the incinerator bed temperature can be held as much as 200°F higher than the laboratory determined fusion temperature without adversely affecting the operation.

Fluid Bed Performance

The unit was designed based on a superficial space velocity through the bed of 3 ft/sec with a bed particle size distribution similar to that represented by curve A, Figure 9.2. It was anticipated that the particle size would tend to

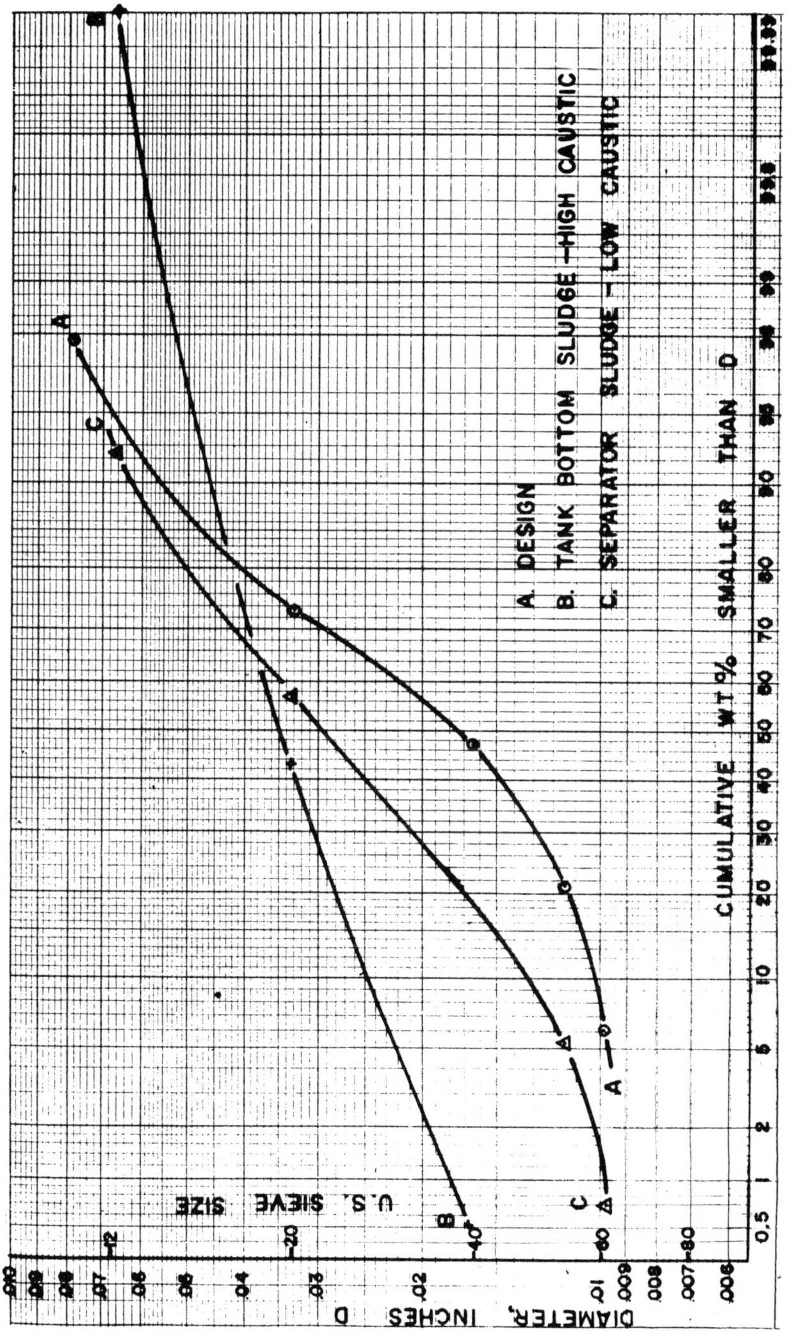

FIGURE 9.2: BED PARTICLE SIZE DISTRIBUTION

increase as the dissolved solids from the spent caustic were deposited on the particles in the bed. Provided a sufficient amount of fine material was introduced with the sludge to serve as new nuclei for the deposition of these chemicals as excessive bed material was withdrawn, the bed particle size distribution would remain satisfactory.

However, when it was found necessary to inject caustic and sludge into the bed instead of above the bed, the superficial space velocity increased considerably, to as high as 5 ft/sec thus making it impossible to retain the small particles contributed by the sludge long enough for them to serve as nuclei for the deposition of chemicals. As a result the bed particles become large and relatively uniform in size. A typical size distribution is represented by curve B, Figure 9.2.

In spite of the large mean size and poor size distribution of the bed particles, operations have been sustained for as long as 23 days. The relatively large mean particle size and poor distribution in the Mandan Incinerator fluid bed has produced adverse effects on operations as described below.

Thumping: The Mandan incinerator vessel emits a violent thumping sound. Shock waves are transmitted to the unit foundations. The vibrations are obvious to anyone standing near the vessel. The violence of the thumping increases as the pressure drop across the fluid bed increases. It is speculated that the bed is slugging in contrast to a relatively steady fluidized state.

Initial Fluidization of the Bed During Startup: During a startup after the fluid bed had been permitted to slump and cool, it was not possible to bring the bed to a fluidized state with combustion air. At the time it was suggested that the bed had become moist causing the particles to adhere to each other. The bed was partially removed and replaced with sand and the startup continued.

Bed material with a size distribution represented by Curve B, Figure 9.2, a particle density of 176 lb/ft^3, and an apparent bulk density of 88 lb/ft^3 produced satisfactory operation of the unit with a superficial space velocity of 4.74 ft/sec. The fluidized bed density was 64.4 lb/ft^3. Utilizing the smoothed correlation of particulate fluidization curves, it was determined that the incipient fluidization velocity for this material would have been 2.03 ft/sec. Combustion air at 100°F would have produced a superficial space velocity of only 1.12 ft/sec.

It would have been necessary to bring the combustion air to 565°F before the incipient fluidization velocity would have been reached. While the presence of moisture has not been eliminated as a factor contributing to the startup difficulties desscribed above, it is highly probable that the bed simply required a higher velocity for fluidization than was possible with cool combustion air.

Torch Oil Ignition: The bed temperature at which torch oil ignites has been unpredictable. Early considerations led to prescribing a minimum bed temperature of 900°F to be reached before injecting torch oil. With the initial sand fill, this temperature proved to be adequate. However during subsequent startups when the unit contained a chemical bed from the previous run, torch oil would not ignite at 900°F. With a 900°F fluid bed temperature and over 10% oxygen in the flue gas, the injection of torch oil would produce a billowing yellow cloud of smoke from the stack with no effect on bed temperature. It was necessary to raise the bed temperature to 1050°F before the torch oil would ignite.

This problem has been attributed to poor fluidization of the bed as a result of a large particle size with inadequate space velocity for proper fluidization. Apparently large bubble-like slugs of air and vaporized torch oil pass through the bed. Ignition is inhibited by poor heat transfer from the bed particles. When the average particle size is reduced by making up part of the startup bed with fine sand, the torch oil ignites without difficulty at 900°F.

Bed Particle Size Growth: When spent caustic is injected into the incinerator, the water evaporates, combustible materials are oxidized, the alkaline materials react with the acidic gases from combustion, and the solids thus produced are deposited on the surface of the fluid bed particles causing them to increase in size. Bed particle size growth may be offset by the introduction of smaller particles along with the sludge and by natural attrition occurring in the bed. However, in spite of these offsetting actions, often the bed particle size has been observed to increase quite rapidly ultimately causing a complete loss of bed fluidity. Attempts to reduce the particle size by jetting steam into the bed has met with limited success.

Particle growth rate is related to the mass of the bed and the rate of mass addition. The rate of particle growth is also directly proportional to the diameter of the particles in the bed. The average particle size, therefore, must be controlled to keep the growth rate at or below the rate of particle size reduction by attrition and fines additions. When attempts are made to reduce the average particle size by injecting steam into the bed, there is an apparent improvement. However, because of the loss of fines through an inefficient cyclone with an inadequate system for returning fines and a less than adequate method of steam attrition, the average particle size in the bed cannot be reduced enough to halt the spiraling effect of particle diameter on growth rate and fluidity is ultimately lost.

Loss of Fluidization: The interruption of operations is most frequently caused by the loss of bed fluidity. It was recognized that the introduction of spent caustic would cause some particle size growth and some changes in the bed characteristics. However, it was believed that particles introduced with the sludge would serve as new nuclei for the deposition of chemicals and as excessive bed material was withdrawn a tolerable equilibrium would be reached.

When a bed of finely divided solids is in a fluid state, the fluid bed density decreases as the superficial space velocity is increased. The maximum bed density that can be reached by reducing velocity is represented by the density of the mass of particles at rest. This is called the apparent bulk density (ABD). Since fluidization causes an expansion of the mass of particles, the ratio of fluid density to apparent bulk density serves as a measure of relative fluidity. Fluidity was nearly lost on several occasions, at various fluid densities. The apparent bulk density of the bed must have been changing.

Apparent bulk density is a function of the particle packing configuration and in a system containing uniformly sized particles, it is independent of particle size. In a system containing variously sized particles, the apparent bulk density decreases as uniformity increases the limiting value being the same whether the uniformity approaches the larger or the smaller of the sizes in the distribution.

It was proposed that untimely termination of operations can be avoided by appropriate action if instrumentation is provided to furnish a continuous record of fluid bed density and the relative fluidity is determined periodically by comparing fluid bed density with ABD. (ABD can be simply measured by weighing a known volume of bed material and converting the volume weight relationship to lb/ft^3.) The ability to control the ratio of the rate of fine material additions to the rate of dissolved solids additions is essential.

UNIT PERFORMANCE

Sludge feed rates were observed between 1.8 and 2.9 bbl/hr and caustic rates varied between 0.7 and 2.0 bbl/hr. Heat requirements per gallon of total feed ranged from 25,000 to 39,000.

The ash produced (bed material) is relatively inert. The maximum hydrate alkalinity observed was 1.4 weight percent. Solubility ranges between 50 and 80 weight percent (in hot 212°F water). The possibilities of utilizing this material or reclaiming chemicals from it exist but have not been investigated.

Stack gases contain water vapor, carbon dioxide, oxygen, and nitrogen. Only traces of sulfur dioxide (less than 0.1 ppm) were found. On one occasion, 2.2 volume percent carbon monoxide was detected when the oxygen content was 5%. CO was not evident when the O_2 content was over 8%.

The scrubber effluent water (see Table 1) contains suspended solids, dissolved solids, and chemical oxygen demand. These waste water contaminants would be greatly reduced by improving the cyclone separator efficiency and recirculating scrubber water.

Heat and Material Balance

In estimating the heating value of sludge, the percent oil is determined and its heating value is estimated from the quality of oil recovered. The sediment is ignited and the volatile matter from the sediment is assumed to be oil. A more accurate value of sludge heating value could be established by use of a bomb calorimeter.

TABLE 1: MANDAN FLUID BED INCINERATOR PERFORMANCE DATA

TEST DATE	I 10/20/69	II 11/19/69	III 3/13/70	IV 3/17/70	V 3/23/70	VI 5/25/70
SCRUBBER Off Gas						
ORSAT CO Vol %	2.2	0	0	0	0	0
ORSAT CO_2 Vol %	10.8	4.4	5.4	6.2	4.4	7.0
ORSAT O_2 Vol %	5.0	8.0	11.0	10.0	15.4	11.0
Hydrocarbon, PPM Vol	-	-	0	0	0	0
SO_2, PPM Vol	-	-	0	T	T	T
Particulates Grams/SCF (Dry)	-	-	0.058	0.040	0.046	0.098
Influent Water						
pH	6.9	6.9	7.0	6.7	6.4	7.8
Suspended Solids	24.0	30.0	36.0	18.0	26.0	58.0
Dissolved Solids	764.0	938.0	1108.0	1054.0	934.0	1148.0
COD	72.0	64.0	56.0	168.0	52.0	124.0
Effluent Water						
pH	6.0	6.0	6.9	6.6	6.5	7.5
Suspended Solids	722.0	530.0	321.0	207.0	283.0	2406.0
Dissolved Solids	1169.0	1628.0	2092.0	1394.0	1564.0	1792.0
COD	144.0	160.0	28.0	60.0	36.0	312.0

OPERATING AND MAINTENANCE COSTS

Automatic controls are provided to allow the unit to operate with minimum attendance by an operator, and to shut down the unit in the event of a hazardous malfunction. Under normal operations, approximately 25% of an operator's time is required for this unit. When tank bottoms are being transported to the unit, the number of man hours required may increase appreciably.

Utilities required for the unit appear reasonable. The connected electric load is approximately 48.2 hp per ton of sludge feed, most of which is required for the air blower. Natural gas and torch oil are required for startup only. Compressed air (100 psig) at 13,000 SCFH, steam (140 psig) at 900 lb/hr, and raw water usage at approximately 35,000 lb/hr are the only other utilities required.

Installation of piping to facilitate using raw water in place of treated cooling water has reduced water costs approximately 70%. Maintenance costs will average approximately 1.7% per year of installed cost per year. The two major contributors to maintenance cost are feed nozzle plugging and pump wear caused by the abrasiveness of the sludge. The following is a breakdown of projected operating costs based on a yearly operating factor of 33.2%.

Operating	Maintenance	Utilities	Total
$4,813	$5,854*	$6,653**	$17,320

*Maintenance costs are based on average expenditures for quarter ending June 30, 1970. These initial costs are high for this prototype unit, and will undoubtedly decrease in the future.

**The water costs included in the utilities figure are based on using untreated water.

A yearly operating cost of $17,320, processing 8,712 bbl/yr of waste caustic and sludge, results in a unit cost of 4.7¢/gallon of waste. The Mandan Refinery has a crude running capacity of 50,000 bbl/day. Based on crude running capacity, this cost is 0.095¢ per bbl of crude processed. It is noteworthy however, that the Mandan Refinery processes a low sulfur (0.4 wt percent) 42° API gravity crude containing little residual fuels or asphaltic material. The production of spent caustic and oily sludge is therefore at a minimum compared with the refining industry.

CONCLUSIONS

(1) The fluid bed incineration process has been demonstrated to be practical and effective for the disposal of petroleum refinery generated spent caustic and oily sludge.

(2) The process creates no atmospheric pollution problems, emitting only carbon dioxide nitrogen and water vapor. The odor of the off gas has been described by various observers as being slight to nonexistent.

(3) The ash produced contains sodium carbonate, sodium sulfate, other soluble inorganic salts and inert material such as sand, clay, rust, etc.

(4) Operations are frequently prematurely terminated by the loss of bed fluidity caused by excessive bed particle size. The particle size growth rate can be controlled by including salient design features and by exercising proper operating techniques.

RECOMMENDATIONS

The Mandan Refinery fluid bed incinerator is functional. Run lengths are limited by the loss of bed fluidity resulting from excessive average particle size and poor particle size distribution in the bed. The design of future units should include the following considerations.

Superficial Space Velocity: Avoid excessive elutriation. Provide sufficient air capacity for fluidization of bed when starting up cold.

Dust Recovery: An efficient dust recovery system that continuously returns fines to the bed is necessary to provide small nuclei on which to deposit the chemicals thus avoiding an excessive particle growth rate. Bed material lost through the dust recovery system is scrubbed out of the off gasses with water. The waste water from the scrubber is contaminated with suspended and dissolved solids from the unrecovered fines.

Attrition or Fines Recirculation System: An efficient easily controlled attrition or fines recirculation system must be provided so that an uncontrolled particle growth rate can be avoided.

Instrumentation: Continuous recording pressure differential instruments should be provided across the total fluid bed and across part of the fluid bed so that operators can control bed depth and observe bed density.

Operation: Timely knowledge of the relative fluidity (fluid bed density/apparent bulk density) is essential if appropriate action is to be taken to avoid the complete loss of fluidity.

Feed Storage and Insertion : Spent caustic and oily sludge should be stored separately and independently fed with rate controls system provided so that the unit heat balance and particle size growth rate may be kept under control.

Fluidized Bed Incineration of Carbonaceous Industrial Wastes

The material presented in this chapter is based on EPA PB 211,161 March 1972 prepared by the Battelle Laboratories for the State of Ohio Department of Natural Resources.

The disposal of carbonaceous liquid wastes by fluidized bed incineration usually consists of the following steps:

(1) Concentrating the wastes in multiple effect evaporators to a level where the fuel value of the concentrate is sufficient to evaporate the remaining water and raise the temperature of the waste to the desired incineration temperature.

(2) Introducing the concentrated waste into the top of the incinerator as a fine spray. However, viscose wastes and wastes containing large amounts of volatile solvents were introduced directly into the bed of fluidized solids.

(3) Contacting the droplets with a bed of fluidized solids held at a temperature sufficiently high to burn all of the carbonaceous matter.

(4) Recovering the residual inorganics and noncombustible portion of the waste either from the bed of fluidized solids as an agglomerated product or from the exhaust gases as a finely divided dust.

The operating conditions in a fluidized bed system can be readily controlled within very narrow ranges, a factor which makes possible the incineration of carbonaceous wastes while minimizing the generation of noxious fumes. In addition, these systems efficiently utilize the fuel value of the wastes. They can be operated at temperatures which will effect the agglomeration and the retention of many materials in the bed of fluidized solids, thus minimizing the entrainment of particulate matter in the exhaust gases, if desired. Because of the effectiveness of fluidized bed incineration systems in the disposal of pulping and papermaking wastes, as well as other carbonaceous wastes, it was believed that such systems also might be applied advantageously to the abatement of water pollution by carbonaceous waste from other industries.

SAMPLES SELECTED FOR EXPERIMENTATION

Prior to undertaking the experimental studies, various waste samples obtained from four industries were evaluated to determine which were most amenable to fluidized bed incineration. The samples were selected, primarily, on the basis of (1) solids content, (2) heating value, (3) pollutional constitutents, and (4) quantity produced. Each of the four industries will be discussed separately.

Paint

One of the wastes selected from the paint industry was the bottoms or sludge remaining after distillation of solvents that are used for cleaning equipment. Two solvent bearing sludges were selected for experimental incineration in the fluidized bed unit. One of the sludges was from a solvent recovery (distillation) operation of a large paint manufacturing plant. It was felt that this sludge was representative of the paint manufacturing plants that operate their own recovery system. This waste contained about 52 percent total solids and 0.8 percent ash. The heating value of the waste, as received, was 16,130 Btu/lb.

The second sludge or still bottoms selected for fluidized bed incineration experiments was from a firm that recovers solvent from several paint plants and does not manufacture paint. This sludge contained about 37.7 percent solids and 6.8 percent ash. The heating value of this material, as received, was 15,910 Btu/lb.

A third waste selected for study was the washout water from latex paint production, which represented a typical aqueous waste generated by the paint industry. The total solids content of this waste was only 7 percent, and its heating value was 3,920 Btu/lb. This waste was selected for experimentation primarily on the basis of its high pollutional potential.

One additional waste from the paint industry was selected for experimentation because it was felt that it could be used as cooling water during incineration of the high heating value still bottoms. This waste was the very dilute (49,000 mg solids/liter) water of reaction (decanter water) collected during resin production.

Plastics

Wastes generated in the production of three different types of plastics were selected for experimentation. The first of these was a sludge from the primary treatment of polyvinyl chloride process waste water. This waste contained 9.7 percent total solids and 1.05 percent ash or inorganic residue. The solids from this waste had a heating value of 5,610 Btu/lb.

A similar waste from the plastics industry selected for experimentation was sludge from the primary treatment of styrene process waste waters. This material contained 9.3 percent total solids and 0.51 percent ash. The heating value of the solids portion of this material was 14,970 Btu/lb.

In addition to the primary treatment sludges, several types of solid and liquid wastes were selected from the production of laminated plastics in the hope that a mixture suitable (pumpable) for feeding to the incinerator could be prepared. These wastes included: (1) sanding dust, (2) resin production reaction water, (3) waste solvents, and (4) waste resins. The heating value of the dust fraction alone was in excess of 9,000 Btu/lb.

Rubber

On the basis of the characterization and analysis of the samples obtained from the rubber industry two types of waste were selected for experimentation in the fluidized bed incinerator. These were: (1) concentrated digestor wastes from rubber reclaiming operations, and (2) synthetic rubber production sludges after coagulation/flotation. The digestor waste contained 2.5 percent solids and 0.16 percent ash and had a heating value of 9,820 Btu/lb. The sludge from synthetic rubber production contained about 15.4 percent solids and 8.8 percent ash. The heating value of the dried solids was 14,950 Btu/lb.

Textiles

The only waste materials from the textile industry that appeared to be suitable for fluidized bed incineration was the waste from viscose rayon production. The viscose rayon waste contained 23 percent solids and about 9 percent ash. The heating value of the solids was 4,415 Btu/lb.

DESCRIPTION OF EQUIPMENT

A detailed sketch of the fluidized bed incinerator unit and auxiliary equipment which was used during this program is shown in Figure 10.1. The fluidized bed incinerator had an inside diameter of 10 inches in the bed zone. The freeboard area tapered from 10 inches diameter at the bed level to 16 inches diameter at the gas exhaust port, with an overall height of 72 inches. The unit was lined with 6 inches of castable insulating refractory and had a ⅜ inch thick stainless steel orifice plate containing thirteen ¼ inch diameter orifice tubes. The orifice tubes were constructed in such a manner (tee types) as to prevent the bed material from running into the windbox during shut-down.

The fluidized bed unit was mounted on a windbox combustion chamber in which natural gas was burned and mixed with excess air to provide a mixture of gases at a controlled temperature. These gases passed through the orifice plate to fluidize the solid bed material and in some cases provided a portion of the heat required for incineration of the wastes. The dust laden exhaust gases from the bed were discharged through a 2 inch opening in the top of the unit into a cyclone dust collection system. After the cyclone, the gases passed through a Venturi-type scrubber and then through an entrainment separator.

FIGURE 10.1: SKETCH OF 10 INCH DIAMETER FLUIDIZED BED UNIT

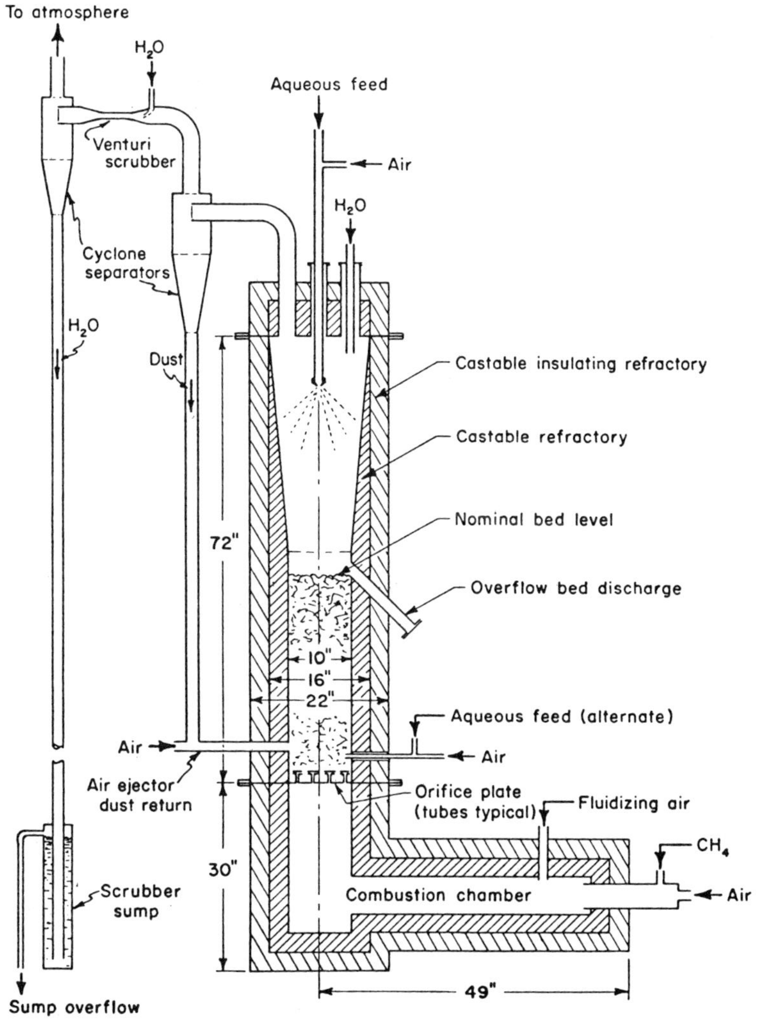

The mixture of solids and liquid from the entrainment separator flowed into a sump which provided a water seal for the slurry discharge system. The effluent from this sump was overflowed to waste or recycled to the scrubber. The gases emerging from the entrainment separator were exhausted from the building through a 2 inch line.

Thermocouples and pressure taps were placed at strategic points in the system to measure temperatures of inlet gases, bed material, freeboard space, and off-gas stream; and pressure differentials across the entire system, the bed zone, and the dust collection system. The temperature of the bed was controlled during incineration by varying the flow of natural gas and/or the rate of introduction of the particular waste material into the unit.

The waste material was pumped to a pneumatic feed gun mounted on the side of the incinerator using a pump. The feed gun was a single barrel type discharging in the bed zone below the top of the bed. The waste material was dispersed into droplets of the desired size by varying the amount of air to this gun. The solid residue, or ash, from the waste material was discharged from the incinerator as dust from the cyclone.

DISCUSSION OF EXPERIMENTAL RESULTS

Paint

The results of the experimental work conducted on the fluidized bed incineration of wastes from the manufacture of paint indicate that this type of incineration is technically feasible. The physical operation of the incineration system was quite good and very few operational problems were encountered. Early in the program, it was found that introduction of wastes into the incinerator under the bed rather than as an overhead spray gave improved results. This was due to the high volatile organic content of the wastes. Introduction of these wastes as an overhead spray did not afford sufficient retention time for complete combustion of the volatiles and organics.

During the incineration of the sludges from solvent recovery operations, it was necessary to use cooling water in the freeboard of the incinerator to prevent excessively high temperatures in this zone. This is due, primarily, to the high heating value of these wastes. In one of the experimental runs water of reaction from resin production was used as cooling water. This appears to be quite attractive, since two wastes are treated in one operation.

The incineration of the sludges from solvent recovery can be carried out autogenously, that is, without the use of auxiliary fuel, because of the high heating value of these wastes. In the incineration of the wash-out water from latex paint production fuel is required to elevate the temperature to a point where combustion will occur.

Plastics

The experimentation on the fluidized bed incineration of wastes from plastic production indicated that this too is technically feasible. The wastes incinerated experimentally, sludges from primary treatment of styrene and polyvinyl chloride process wastes, are amenable to fluidized bed incineration. Even though these materials require auxiliary fuel for combustion, this method of incineration remains quite attractive since little or no noxious or toxic gases are produced.

The residue from the incineration of these wastes is a finely divided dust that could be disposed of in a landfill. They contain no organic compounds that could be leached out and cause stream pollution problems. There also is the possibility of recovering the calcium, iron, aluminum, or magnesium from this residue prior to disposal.

The only effluent stream from this type of incineration that might cause stream pollution is the scrubber effluent. However, in a commercial installation of this type the scrubbing water could be recycled to the scrubber to alleviate this problem.

Rubber

The results obtained from the experimentation on fluidized bed incineration of wastes from rubber manufacturing indicate the technical feasibility of the technique for treating the wastes under consideration. The only waste incinerated from this area, however, was the flotation sludge from the primary treatment of synthetic rubber production wastes.

The sludge from synthetic rubber production was incinerated under conditions which did not produce toxic or noxious gases. The effluent from the exhaust gas scrubber contained only trace contaminants. When secondary sewage effluent was used as scrubbing media, the COD level increased from 36 to 211 milligrams per liter. The mercury level increased during the run from 0.002 to 0.012 ppm. The physical operation of the incineration system was quite good during these experimental runs.

Textiles

The only waste procured from the textile industry that appeared suitable for incineration was the waste from a viscose process. From the data collected during the experimentation it would appear that complete combustion of the organics present in this waste is not possible by this technique. It is, however, possible to incinerate this particularly difficult waste material and achieve about 80 percent combustion. This would alleviate the potential pollution problem that now exists from disposal of this material in a landfill.

Complete combustion could not be attained because of the low melting point of this material requiring that the incinerator be operated at low temperatures.

The low melting point is due to the high caustic content of this waste. Attempts to operate the incinerator at temperatures above this point caused the bed material (sand) to become sticky and defluidize. However, even though complete combustion was not achieved, partial incineration does not produce toxic or noxious gases.

POTENTIAL IMPACT ON WATER EFFLUENTS

Paint

The results of the experimental program indicate that in the case of solvent bearing wastes from the paint industry, the total waste load can be reduced considerably by fluidized bed incineration. The average total waste remaining for disposal after incineration of solvent recovery sludges is about 3.5 percent of the original weight.

The average chemical oxygen demand of this type waste is reduced from 547,250 mg/l to essentially zero. However, the effluent from the scrubber used for cleaning the exhaust gases from the incinerator contains an average of about 24 mg/l COD. This effluent could be recycled to the scrubber continuously and incinerated concurrently with the solvent bearing wastes.

On the basis of a typical plant which generates about 125,000 gallons per year of solvent recovery sludge, the use of incineration would reduce COD loading to the environment by about 525,000 pounds per year for this plant. Assuming the plant to be representative of the national picture, a total reduction of 360,000,000 pounds of COD per year could be achieved in the United States.

Plastics

Incineration of the primary treatment sludges reduces the waste to be disposed of to less than 5 percent of the original weight. The residue from the liquid wastes would amount to only about 75 mg/l and the residue from the solid waste would amount to only about 2 percent or less if they were incinerated. As was the case with incineration of the wastes from paint manufacture, the only COD in the products occurs in the scrubber effluent, which amounts to only about 8 mg/l and this could be recycled and incinerated also.

Using as a basis a typical plant which generates 6,000,000 pounds per year of primary treatment sludge, the possible reduction in COD by incineration is about 3,500,000 pounds per year. Assuming this plant to be representative of the entire plastics industry nationally which consists of 340 plants, the total COD reduction for the United States would be about 1,200,000,000 pounds per year.

Rubber

On the basis of one large synthetic rubber manufacturing plant which provided quantities of flotation sludge for experimental studies, a possible reduction in COD by incineration was calculated at about 1,000,000 pounds annually. Projecting this to a national level which consists of only 28 plants, the total COD reduction would be about 88,000,000 pounds.

Textiles

Incineration of the dilute dye wastes would not be economically feasible because of the large quantities of water involved. Incineration of the viscose process waste should be considered because of the high COD content of this material. However, additional research is needed for the incineration of this particular waste to develop the proper operating parameters. No information was obtained on the quantities of viscose waste generated, but the amount probably is low since there are only 20 plants producing viscose in the United States.

PRELIMINARY ECONOMIC EVALUATION

To assess the economic feasibility of incineration for disposal of wastes in the paint, plastics, rubber, and textile industries, a preliminary evaluation was made of the process requirements and costs for three specific incineration systems. These systems were selected on the basis of experimental results and conclusions which indicated:

(1) The technical feasibility of incineration of waste solvents and sludges from the paint industry using the heating value of these wastes as a prime source of energy. The envisaged process also would permit the incineration of certain other paint wastes such as reaction water and latex wash-out concurrently with the solvent bearing wastes. Flows would be adjusted so that autogenous operation could be achieved.

(2) The technical feasibility of incineration of sludges from plastic primary treatment operations. These sludges could contain either styrene or PVC polymers. On the basis of experimental results, it was determined that additional energy input would be needed for incineration of these sludges. This could be supplied by burning methane and/or fuel oil in a combustion zone or by using solid plastic wastes as a source of energy.

(3) The incineration of rubber manufacturing waste sludges with additional energy being supplied by combustion of methane and/or fuel oil.

The following are design aspects of an incinerator which would be required for each of these three systems.

Incineration Processes

Paint Waste – The envisaged process for the incineration of solvent recovery sludges, reaction waters, and latex wastes is illustrated in Figure 10.2.

FIGURE 10.2: SIMPLIFIED FLOWSHEET OF INCINERATOR FOR PAINT WASTES

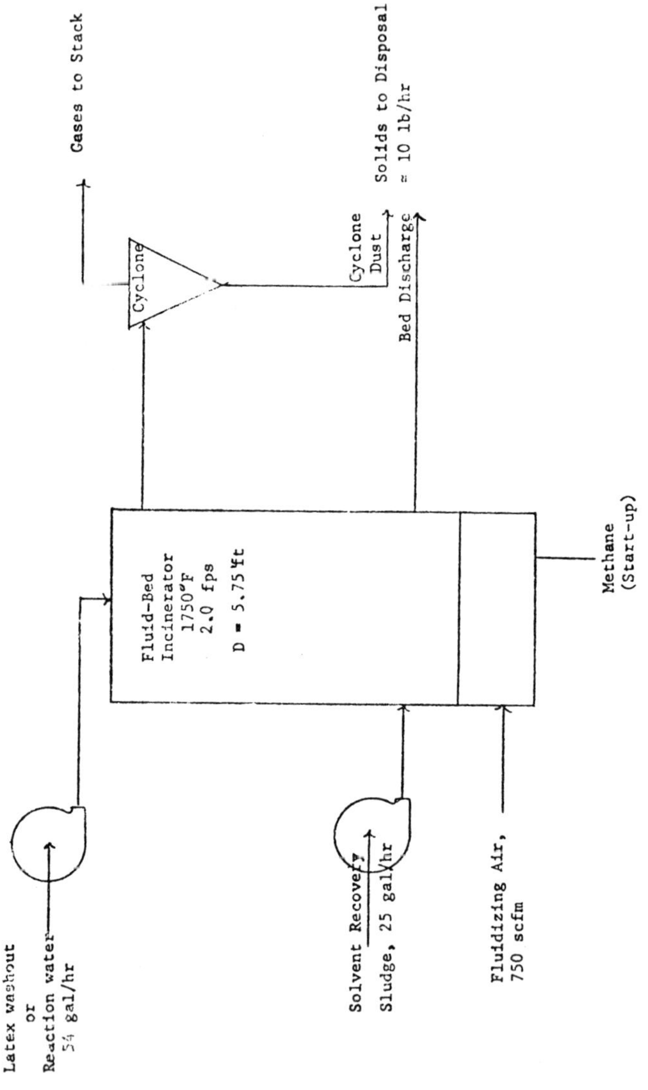

Capacity was based on the production of 125,000 gallons annually of solvent recovery sludge which had an average heating value of 120,000 Btu/gal. Operating conditions were selected on the basis of experimental runs made during the study, (i.e., bed temperature 1750°F, superficial velocity 2.0 fps). Calculations made to determine material and energy balances indicate that an incinerator of about 6 feet in diameter would be required for this system. The results also indicate that approximately 54 gal/hr of latex wastes and/or reaction water could be disposed of by incineration concurrently with the waste sludge.

Plastic Waste – The envisaged process for the incineration of primary treatment sludge from plastic manufacture is shown in Figure 10.3. Capacity was based on the production of 30,000,000 pounds annually of sludge containing an average of 10 percent solids whose heating value was estimated at 15,000 Btu/lb. Operating conditions were selected on the basis of experimental runs which indicated a bed temperature of 1325°F and superficial velocity of 1.8 fps were required. Calculations made to determine material and energy requirements indicate that an incinerator of about 12 feet in diameter would be required for incineration of this waste sludge. Added fuel must be supplied, (i.e., 156 scfm of methane or 625 pounds per hour of waste polymers).

Rubber Waste – A simplified flowsheet for incineration of a flotation sludge containing rubber wastes is shown in Figure 10.4. In this example system, no information was obtainable for a typical quantity of wastes generated by plants. Thus, capacity was arbitrarily selected at 10,000,000 pounds annually. A typical sludge, however, is known to contain 15.4 percent solids which have a heating value of about 15,000 Btu/lb of dry solids. Operating conditions from experimental runs with this sludge would be bed temperature of 1400°F and superficial velocity of 2.0 fps. The results of design calculations indicate this incinerator would be about 7.5 feet in diameter with provision for burning about 32 scfm of methane in a combustion chamber.

Estimated Capital and Operating Costs

The essential capital and operating costs for the three systems selected above are itemized in Table 1. Basically, the capital costs were estimated for conventional fluidized bed units currently being manufactured by a number of companies. Operating costs were selected for the major items that were believed to be significant based on experience with these types of systems.

Generally, the cost data indicate that fluidized bed incineration could be used for disposal of these wastes at an operating cost of between 140 to 180 dollars per day for plants generating between 3,000,000 and 30,000,000 pounds of waste annually. This reduces to a cost of between 0.1 to 1.0 cent per pound of waste being disposed of by incineration. The higher cost, for paint waste incineration, results from the use of a small diameter fluidized bed unit and from the use of low feed rates necessary for the high heating value of these wastes.

FIGURE 10.3: SIMPLIFIED FLOWSHEET OF INCINERATOR FOR PLASTIC WASTES

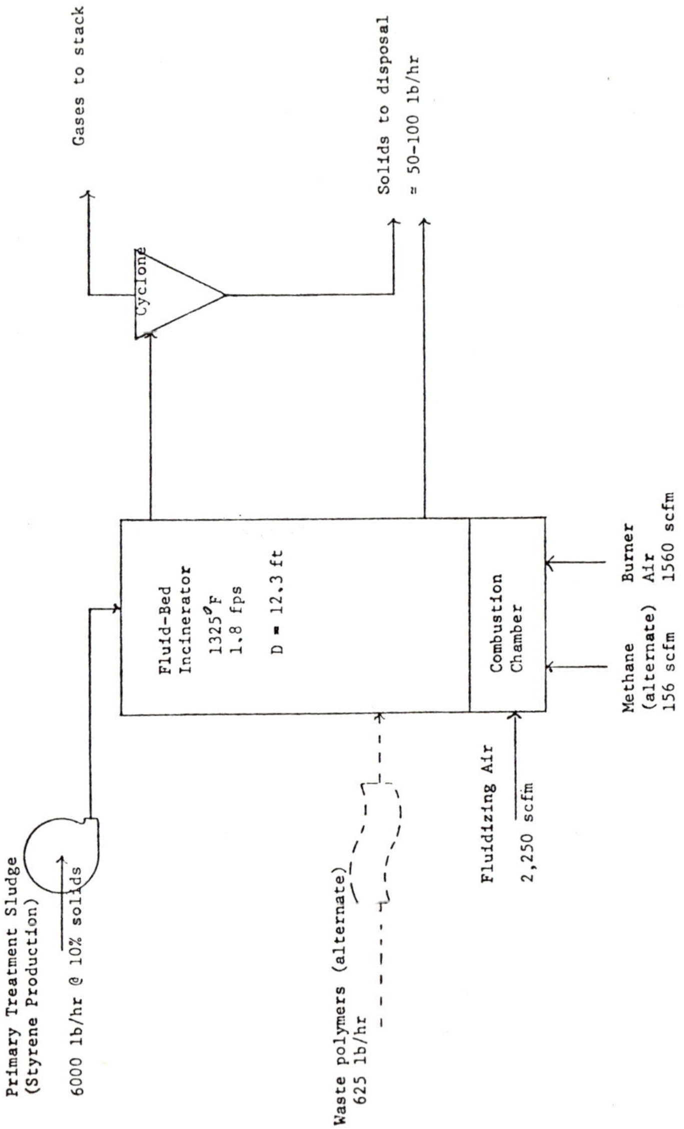

FIGURE 10.4: SIMPLIFIED FLOWSHEET OF INCINERATOR FOR RUBBER WASTES

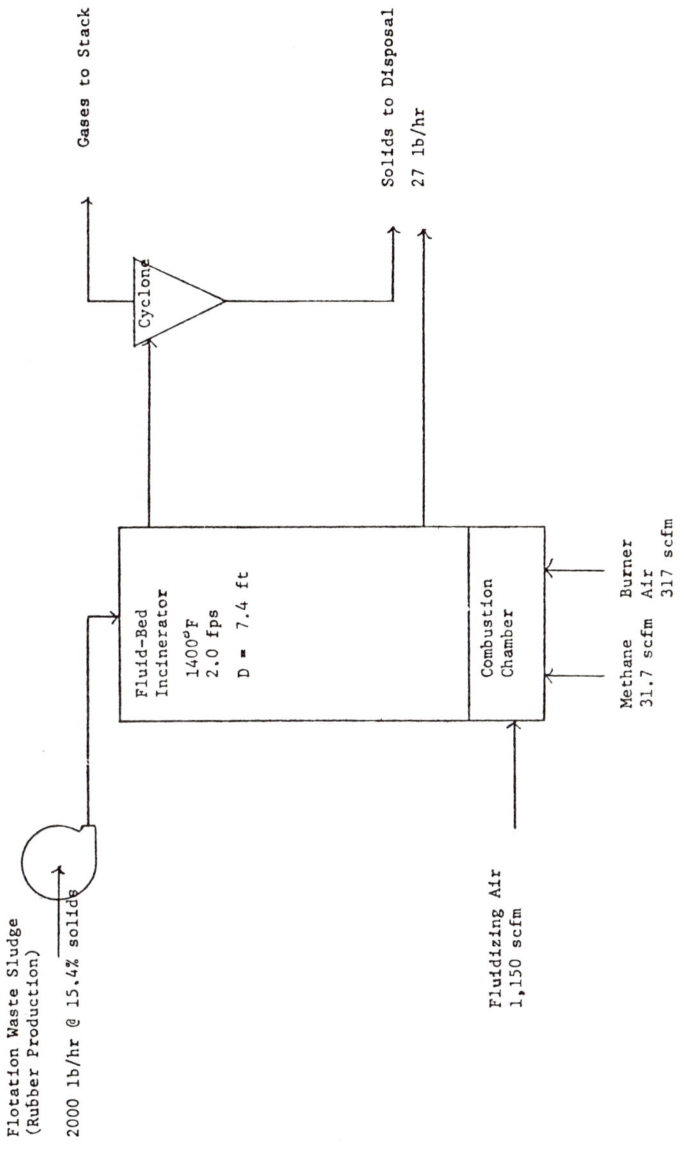

TABLE 1: ESTIMATED CAPITAL AND OPERATING COST FOR INCINERATION OF VARIOUS WASTES

	Paint	Plastic	Rubber
Essential Plant Cost, $	$200,000	$350,000	$250,000
Essential Operating Costs, $/day			
1. Fuel @ $.25 per 10^6 Btu	0	0	4.10
2. Electric Power @ $.007/KWH	8.40	8.40	8.40
3. Supplies and Maintenance @ 0.003% of Plant Cost	6.00	10.50	7.50
4. Operating Labor @ 24 Man-hr per day, $3.50 per Man-hr	84.00	84.00	84.00
5. Amortization @ 0.0224% of Plant Cost	44.80	78.50	56.00
Total Operating Costs	143.20	181.40	160.00
Cost per Pound of Waste	1.1¢	0.14¢	0.40¢

CONCLUSIONS

This study has indicated that certain waste materials generated by the paint, plastic, rubber, and textile industries can be successfully disposed of by incineration in a fluidized bed system. Based on a survey of these industries in Ohio and on the results of experimental incineration studies, the following conclusions were drawn.

(1) In the paint industry, an incineration process would be particularly effective for disposal of solvent recovery sludges and other aqueous wastes such as latex wash-out water and reaction waters from resin production. These wastes can create difficulties by disposal in landfills or municipal sewers. It was experimentally demonstrated that all of these materials can be incinerated concurrently in a fluidized bed system.

No fuel was required due to the high heating value of the solvent bearing sludges, and essentially no noxious or toxic materials were present in either the exhaust gases, the gas scrubber effluent, or the residue remaining after incineration. A full scale incinerator required for a typical plant was estimated to be about 6 feet in diameter and cost about $200,000. Operating costs were estimated at $140 per day or 1.1¢ per pound of waste.

(2) Primary treatment sludges from the plastic industry can be disposed of by fluidized bed incineration using supplemental fuel addition. One source of fuel might be the large quantities of solid plastic waste generated by this industry.

The experimental studies indicate that essentially no toxic or noxious materials are produced during incineration of styrene or PVC wastes other than hydrogen chloride which is removed in gas scrubbing effluent. An incineration facility capable of handling the primary treatment sludge from a typical plant was estimated to be 12 feet in diameter and to cost about $350,000. Operating costs, however, were estimated to be less than 0.2¢ per pound of sludge containing an average of 10 percent solids. Although these wastes currently are being disposed of by landfill and apparently do not represent a serious pollutional danger, it was concluded that incineration might show an economic incentive compared to the cost of disposal by landfill.

(3) Sludges from the rubber industry can be incinerated in a manner similar to that used for plastic wastes. Although very little definite information was obtained, the major sludges that appear amenable to incineration are those from primary treatment of process waste water and those from rubber reclaiming operations. Experimental work with a flotation sludge from a large rubber company indicates excellent incinerator operation and no generation of noxious components. On the basis of 10,000,000 pounds of sludge annually, it was estimated that a 7.5 foot diameter incinerator would be required. Capital costs were estimated to be $250,000 and operating costs of 0.4¢ per pound of sludge incinerated.

(4) The use of incineration for disposal of textile wastes does not appear promising. The size of the textile industry in Ohio has decreased greatly in the last 20 years and probably generates a very low pollutional effect compared to the other industries studied. The only textile waste investigated experimentally in this program was from viscose rayon production; incinerator operation in this case was poor due to the presence of low melting point inorganic waste materials. Extensive additional study would be needed to perfect an incinerator for this waste. Although other wastes are generated by the textile industry, these are generally extremely dilute and not amenable to incineration.

The Melt-Zit High Temperature Incinerator

The material in this chapter is based on a report prepared for the city of Brockton by E.R. Kaiser and published by the U.S. Department of Health, Education, and Welfare as PB 217,062 August 1968.

High temperature incineration as utilized in the Melt-Zit process includes the use of a vertical shaft furnace operating between 2600° and 3200°F. Combustibles are incinerated and noncombustibles are converted to a molten residue which flows as a liquid stream from the base of the unit. When cooled or quenched rapidly in water this residue takes the form of frit or sand-like particles, black in color, sterile and inert.

Recognizing the potentials of high temperature incineration, a test facility was constructed in Massachusetts. This experimental unit operated as a development facility to provide data for commercial designs and operating techniques. It was not intended as a commercial unit. It has requently been displayed in operation in an effort to introduce the high temperature incineration concept and the eventual marketing of the process.

The city of Brockton, Massachusetts, (population 90,000) had completed a broad study of methods of refuse disposal and was impressed with the possibilities of high temperature incineration. The results of tests which were conducted in March 1968 are described in detail in the original report. Test results of particular interest to public officials considering incineration are evaluated here.

THE MELT-ZIT FACILITY

The Melt-Zit principle is unique in the art of incineration. The noncombustible fractions of the refuse are melted in a bed of high temperature coke, and drained from the furnace as molten slag and iron. Organic matter in the residue is thereby automatically prevented and complete sterility is achieved. The residue has a high density.

The Melt-Zit pilot installation is basically a verical, cylindrical shaft furnace with refractory lining. The inside diameter is 3 feet (minimum) and the height 55 feet. The molten residue, a slag-like material, including metal, is drained through an opening at the base, above the hearth; it drops into a quenching tank of water where rapid cooling results in a black granulate and iron pellets.

Refuse is charged midway up the stack through an open charging chute. The refuse feed is continuous. The combustible materials burn in suspension while the heavier noncombustibles burn and melt on the surface of the fuel bed.

PERFORMANCE OF MELT-ZIT INCINERATOR

Based on the best period of the tests, the capacity of refuse input for the 3 foot diameter furnace was 1.6 tons per hour. The refuse was municipal rubbish and garbage, 90 percent of the former and 10 percent of the latter, by wieght. The operating temperatures in the coke bed were from 2600° to 3200°F.

The weight of slag (including metal) produced during the longest test run was 460 pounds per ton of refuse, or 23 percent of the weight and 4 percent of the volume of the refuse. The pilot furnace was operated with direct discharge to the atmosphere at the top. The stack gas scrubber with which the furnace was equipped was not used, as it would have prevented proper sampling of the gases.

The maximum length of uninterrupted run during the tests was 62 hours. Interruptions were caused by mechanical failures in the slag removing conveyor system and trouble with the flow of slag due to its high viscosity.

EVALUATION OF MELT-ZIT INCINERATOR

In the state of development and method of operation of the Melt-Zit unit during the March 1968 tests, the pilot incinerator did not perform satisfactorily or reliably. The Melt-Zit incinerator, therefore, needs further development prior to full scale production. It is believed that the process has sufficient promise to warrant further design and development. When improved performance can be demonstrated, the rated tonnages can be increased, probably up to 25 percent.

A full scale incinerator of this type will require a secondary combustion chamber to further burn the gases and the particulate matter. The gases could then be cleaned of dust by conventional dust collectors. It is estimated that the largest practicable size furnace is approximately 8 feet diameter with with 15 tons per hour burning capacity. The technical advantages of the Melt-Zit process are a residue free of putrescible matter, maximum density in landfill, and the elimination of ground water or stream pollution from deposit of incinerator residue.

The most favorable rate of coke consumption during the test period was one ton of coke to 8 tons of refuse. It is believed, however, that by application of suggested improvements a ratio of 1 to 20 could be achieved. Comparative construction and operating costs for a full scale Melt-Zit plant versus a conventional plant cannot be reliably anticipated from the test experience.

Melt-Zit costs include coke and limestone which are not common to other incinerators. These additional costs may be offset, however, by additional benefits such as less area required for landfill and reduced costs of residue disposal.

DESCRIPTION OF MELT-ZIT DESTRUCTOR AND OPERATION

The Melt-Zit pilot scale incinerator is a complete system for refuse reduction, including the following major components:

a. Refuse receiving room
b. Refuse conveyor from receiving room to furnace
c. Coke feeder, coke and limestone elevator to furnace
d. Vertical shaft furnace
e. Venturi flue gas scrubber with refractory lined supply duct (secondary combustion chamber), quench section, flooded elbow, settling tank and sediment conveyor
f. Induced draft fan and flow damper
g. Residue quench tank, residue conveyor, quench water cooler
h. Forced air blowers
i. Control station
j. Building and utilities

Figure 11.1 is a sketch of the equipment that was part of the installation but was not included in the test.

For testing purposes, the system was modified to discharge the furnace exhaust gases directly to the atmosphere without the use of the hot gas flue and gas scrubber. This change was necessary for the planned sampling of the gases for solids. Hence, the test was concerned primarily with the performance of the refractory lined shaft furnace. The possible benefits of the additional combustion space, flow control, and air pollution control were thus not available to the test.

FIGURE 11.1: HOT GAS FLUE AND GAS SCRUBBER

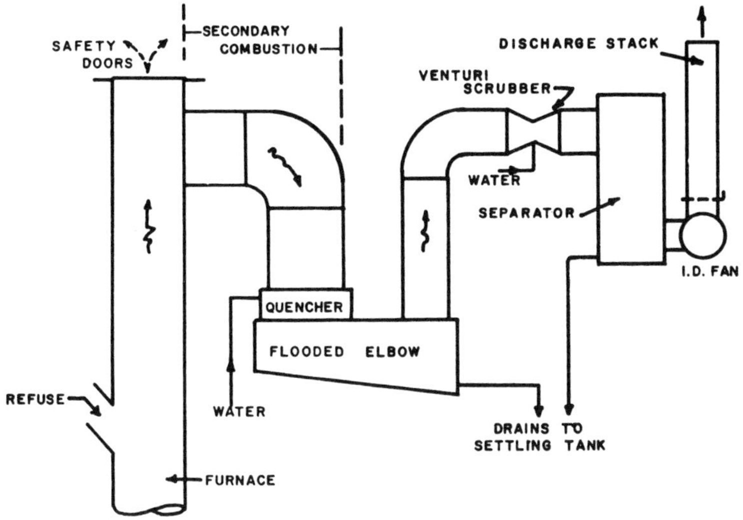

Vertical cross section Figure 11.2 shows the features of the pilot furnace, its several internal diameters, and the means for charging refuse, coke, limestone and air. The feedstocks fall in a random pattern with fairly uniform distribution. Each is charged separately as required by furnace performance.

Air is supplied to the windbox by means of blowers, from which the air flows into the furnace at two levels via air ports spaced around the furnace. Air at the lower level burns the coke and produces the high temperatures for melting metal and slag. Forced air supplied above the coke bed burns the refuse. The furnace produces a slight negative pressure at the feed chute which induces additional air through the charging opening.

As the refuse falls into the shaft, the upward flow of hot gases causes partial entrainment of the solids. Hence, the refuse burns partly on the surface of the coke bed and partly in suspension in the shaft. The pilot furnace was open to the sky for the test, but was surmounted by a screen or grating to limit the size of particles emitted and to control the draft.

Slag and metal are drained through a tap hole at the level of a sand hearth. The molten residue flows into water, where it congeals into a black granulate (frit) and iron pellets.

After the refuse charging rate has been established, coke is charged periodically to maintain the desired level of the coke bed, as observed through a sight glass. Refuse feeding is halted for 5 minutes every 30 minutes when coke is charged and heated.

FIGURE 11.2: MELT-ZIT INCINERATOR, VERTICAL CROSS SECTION

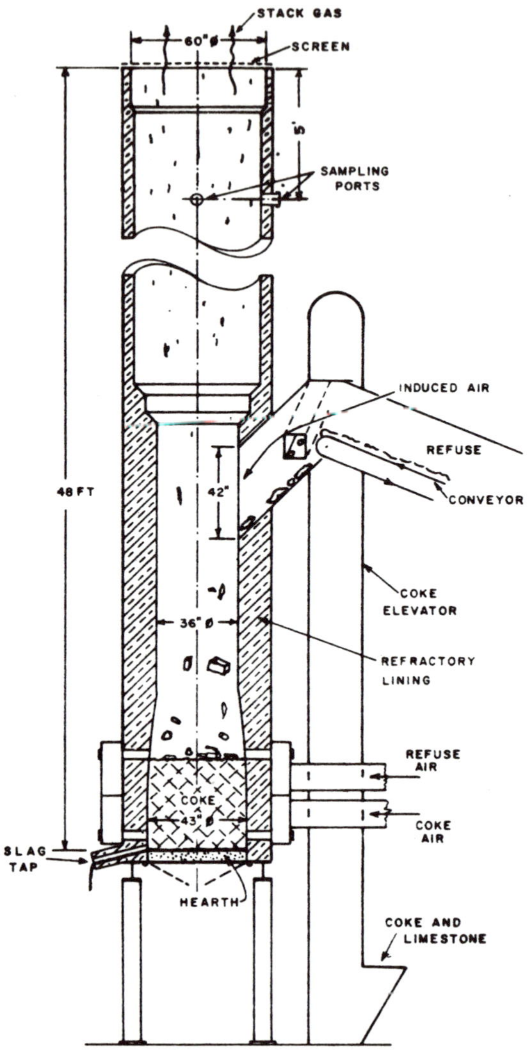

The rapidity of ignition and combustion of the refuse are mutually assisted by the temperature of the shaft refractory. The radiation from the flaming refuse heats the refractory lining. After coke is charged and the top of the coke bed ceases to radiate much heat, radiation from the refractory lining reestablishes prompt ignition of the refuse paper. At no time is the refuse layer on the coke allowed to increase to more than a few inches, so as not to unduly divert the flow of gases from the coke bed. At the end of a run the bottom of the furnace is opened and the remaining coke, slag and molten metal are dropped into a steel buggy.

METAL INCLUSIONS IN SLAG

Refuse iron and steel were melted in the Melt-Zit unit and appeared with small loss in the form of rounded pellets in the frit. The molten furnace product was 20 to 23 percent metal. The pellets alone had a specific gravity of 6.77, and a bulk density of 226 pounds per cubic foot.

The pellets were magnetic, file hard, and contained bubbles of gas. The Rockwell C hardness of 61 and the white brittle fracture indicated them to be white iron due to the formation of cementite, Fe_3C, during rapid chilling. The chemical analyses of eight samples taken during the test indicates the average composition given in Table 1.

TABLE 1: AVERAGE COMPOSITION OF RESIDUE METAL

	Percent		Percent
Iron	95.16	Molybdenum	0.02
Carbon, estimated	2.0	Nickel	0.2
Aluminum	0.03	Silicon	1.6
Chromium	0.3	Titanium	0.06
Copper	0.3	Tin	0.02
Manganese	0.20	Sulfur, estimated	0.1
Magnesium	0.01		

Silver was present under 0.01%. Refuse aluminum was almost entirely converted to Al_2O_3 in the furnace and hence became part of the slag. Aluminum is readily oxidized by carbon dioxide, water vapor and free oxygen in a hot coke bed.

The metal is probably of satisfactory quality for melting furnaces in which the small particle size would not be objectionable. Magnetic separation would not be difficult. The yield would be about 4.6 percent or more of the Brockton refuse, depending on the moisture content of the refuse. If not salvaged, the metal would gradually rust in the landfill.

FURNACE AIR EMISSIONS

The Melt-Zit unit was equipped with a refractory lined flue gas offtake duct and a Venturi gas scrubber. To permit proper sampling of the flue gas for particulate loading ahead of the gas scrubber as originally planned, it was necessary to discharge the gases directly to the atmosphere at the top of the furnace. It was recognized that carbonaceous solids and combustible gases would have continued to burn in the refractory lined offtake. Furthermore, it was recognized that dust collectors of various types and efficiencies could be employed to reduce the particulate loadings before the gases would be discharged in commercial and municipal installations.

The top of the furnace was opened and covered with a stainless steel screen. The screen was an expanded metal mesh with diamond shaped 1 x 5⁄16 inch openings and 59 percent free area. The furnace top was 60 inches diameter. The screen did not fit tightly and suffered distortion from the heat. One purpose of the screen was to hold any oversized pieces of paper, etc., carried up by the gases until they were burned.

Another purpose of the screen was to provide a degree of flow control as a means for balancing the total draft and air ingress to the furnace through the refuse chute. This modification was a simulation of the wet cap gas washer that was used on the furnace prior to the Venturi scrubber installation. The test had been planned on the basis of the wet cap.

In general, the furnace emitted a shower of particles of ash and burning matter, sometimes with flames up to 10 feet above the screen. Charred paper, particles of plastics, ash dust, metal foil and even razor blades were discharged.

In a period of maximum refuse burning the resistance of the screen increased by partial plugging, causing a positive pressure at the sampling port. Sampling had to be discontinued at that time. Because refuse feeding was on and off at unpredictable intervals, the stack gases were affected immediately due to the suspension burning. A period of particulate or gas sampling was a composite of on and off refuse feeding.

The data were vital to the development of the material and heat balances and an understanding of the process. The measurement of the particulate emissions and velocities also was indicative of the limitations on burning rate, the need for secondary combustion space, and the burden on the dust collector.

Particulate Emissions

Five determinations of particulate loading of the gases were completed:

Dust Test No.	Date	Time	Mass Emission Rate, Lb/hr	Corrected to 50 % Excess Air, Lb/1000 lb gas
1	3-18-68	12:00-12:45 P.M.	151.0	13.88
2	3-18-68	3:00- 3:20 P.M.	223.5	23.72
3	3-19-68	1:19- 1:50 P.M.	246.0	13.07
4	3-20-68	9:10- 9:56 A.M.	288.0	18.77
5	3-20-68	2:05- 2:26 P.M.	385.2	28.80

The emission rates were for solids in the gases as the gases approached the screen, and were thus not physically affected by the screen. The particulate loading of Dust Test 4 is most representative. Dust Test 5 was immediately following the end of a test run when the rate for coke air was somewhat higher than normal.

The collected particulate matter was dried, analyzed, and sieved in the laboratory, with results as given in Table 2.

TABLE 2: ANALYSES OF PARTICULATE MATTER EMITTED WITH FURNACE EXHAUST

Dust Test No.	1	2	3	4	5
Combustible, %	31.63	31.45	37.90	42.71	30.33
Ash, %	68.37	68.55	62.10	57.29	69.67
Acetone soluble, %	3.13	1.73	12.43	15.17	2.54
U. S. Sieve Size	Retained on Each Sieve Size, percent				
8	4.29	2.55	5.57	7.10	2.81
20	22.24	25.92	18.01	31.00	35.62
60	27.67	41.03	28.33	39.54	42.92
100	13.20	10.52	14.60	9.27	8.77
200	13.24	10.05	13.71	6.25	4.58
325	6.75	4.03	6.84	2.63	1.56
<325	12.61	5.86	12.97	4.21	3.72
	100.00	99.96	100.03	100.00	99.98

The proportions of suspended solids from the coke, limestone and refuse are not known. Nevertheless, the weight loss in Dust Test 4 was equal to 18.3 percent of the weight of dry refuse charged. The ash content of the weight loss was equal to 26.9 percent of the mineral matter, excluding metals, that were present in the refuse, coke and limestone charged.

The acetone-soluble portion indicates the presence of hydrocarbons in the combustible matter from the refuse. The combustible matter was only partially pyrolyzed to char in its transit through the furnace. The lower combustible and acetone-soluble fraction during Dust Test 5 was caused by less refuse firing and more coke burning during a distress period after a test run (Run A).

The sieve analysis of particulates was the result of a rather vigorous Ro-tap machine sieving for 10 minutes. The friable char was undoubtedly broken down somewhat while bouncing on the sieves. Nevertheless, the large size of the material by dust collection standard indicates that high collection efficiency could be expected if this material were to enter a gas scrubber. With secondary combustion to lower the combustible content, the particle sizes and weights would both decrease. Orsat gas analyses obtained during the period of a test run are assembled in Table 3 with the data calculated from the material balances, based on weights over the entire run.

TABLE 3: FURNACE EXIT GAS DATA FROM A TEST RUN

Time	A.M. 8:45	A.M. 9:10	A.M. 9:21	A.M. 9:10-56	A.M. 11:30	P.M. 2:08	From Balance
				Dry volume, percent			
CO_2	8.0	6.7	7.2	7.36	10.75	10.4	6.84
CO	0.1	0.0	0.1	0.10	0.10	0.1	0.10
O_2	12.3	13.5	13.6	13.00	9.90	9.6	13.60
N_2	79.6	79.8	79.1	79.54	79.25	79.9	79.46
Temp. F	1440	920	1140	1081	1275	940	1270
Excess Air, %	140	178	184	161	89	83	182
Volume, cfm	26,220	-	-	-	20,600	-	18,500
Velocity, fpm	1483	-	-	-	1160	-	1062
Mass flow, lb/hr	32,900	-	-	-	28,800	-	20,608

The above data is from period of refuse feeding; the data from the balances include periods when only coke was charged. When refuse is charged, the percent CO_2 is highest. When only coke was burned, as when the refuse feeding was stopped because of difficulty with slag control, 3.2% CO_2 (Orsat) in the exit gases was measured, with 3.0% at times. Normal air supply to the furnace continued during the refuse feed interruption and lowered CO_2 by dilution. Gas analyses from test Run B in Table 4 below are compared with those calculated from heat and material balances, based on weights over the entire run.

TABLE 4: FURNACE EXIT GAS DATA FROM A SECOND TEST RUN

Time	P.M. 6:42	P.M. 6:53	P.M. 7:05	P.M. 7:10	From Balances
			Dry volume, percent		
CO_2	17.75	7.2	11.0	10.0	10.07
CO	0.10	0.1	0.1	0.3	0.18
O_2	3.30	12.9	7.0	8.2	10.07
N_2	78.85	79.8	81.9	81.5	79.68
Temp. F	1950	1350	1870	1970	1660
Excess Air, %	18	157	47	60	92
Volume, cfm	24,500	-	-	-	18,770
Velocity, fpm	1380	-	-	-	1060
Mass flow, lb/hr	25,100	-	-	-	20,608

In view of the wide fluctuations in burning rates during these two runs and the short sampling periods compared to the entire runs, the correlation of measured values with those from the balances is quite satisfactory.

Smoke

During the first three days of the main 5 day test, no blackness was observed in the exhaust gases, but a brownish to whitish cast was observed with an opacity not exceeding No. 3 Ringelmann equivalent. The same condition held during the last two days except for 5 or 6 short intervals daily of black smoke of No. 4 and 5 Ringelmann, but seldom over 1 or 2 consecutive minutes each. The smoke occurred when the screen was loaded, temperatures were high and excess air was low.

Orsat analyses taken during a smoking condition showed a high carbon dioxide (16.0 percent) and carbon monoxide (1.4 percent) content. The excess air at the time was only 12.9 percent. These smoke conditions could be corrected in a secondary combustion chamber.

Sulfur Oxides

Sulfur oxides (SO_2 and SO_3) in the stack gases were determined by the barium sulfate method during 5 periods, resulting in four determinations, and reported as sulfur dioxide in Table 5.

TABLE 5: SULFUR OXIDES IN FURNACE EXIT GAS*

Date	Time	Average Gas Temperature, °F	Average SO_2 Volume, ppm
3/20/68	9:50 AM - 10:30 AM	1352	65.6
	2:07 PM - 2:24 PM	1020	72.4
3/21/68	9:25 AM - 9:31 AM	2005	113.0
3/22/68	8:10 AM - 9:05 AM		
3/22/68	10:02 AM - 10:50 AM	1900	49.9

*Parts per million

The first two results were obtained during Run A, when the excess air was about 182 percent. The hourly volume of dry gas was 298,000 cubic feet. Hence the average SO_2 volume of 69.0 ppm would represent 20.6 cubic feet of 60°F SO_2 per hour, or 3.56 pounds per hour. The sulfur necessary to produce this SO_2 is 1.78 pounds. As the refuse and coke together contained 7.2 pounds, only 25 percent of the sulfur was converted to SO_2 in the stack gas. This confirms an earlier finding for incinerators generally. The remainder of the sulfur will be found in the fly ash and in the tapped residue.

Oxides of Nitrogen

Four NO_x gas samples were taken when the stack gas was consistently above

1900°F. The grab samples were taken and analyzed by following the phenoldisulfonic acid method. The findings are reported as nitrogen dioxide in Table 6.

TABLE 6: NITROGEN OXIDES IN FURNACE EXIT GAS

Time	----- NO_2 Volume, ppm----- Dry Basis	Wet Basis
8:02 AM	105.9	98.0
8:05 AM	118.3	109.4
10:17 AM	105.0	97.1
10:27 AM	68.2	63.1

As the furnace was probably exhausting 200,000 cubic feet of dry gas an hour at the time, 80 percent nitrogen, the findings indicate that 13.6 to 23.7 cubic feet of NO_2 were generated per hour. The concentrations of NO_x from the Melt-Zit unit were similar to those reported for other municipal Incinerators and much less than those for large industrial boilers or automobiles.

Odor

The sampled gases were regularly checked for odor as they were discharged to atmosphere from the dust sampling apparatus. At no time during any of the five dustloading tests was there any detectable odor in the sampled gases.

The gas temperatures at the time of sampling ranged form 820° to 1320°F, which would ordinarily not destroy all odors. Nevertheless, other experience indicates that when good combustion prevails, as with incinerators that generate 1500° to 1800°F and have secondary combustion chambers, odor from stack gases is negligible.

REFRACTORY SERVICE

The shaft furnace consisted of a cylindrical shell lined with firebrick shapes to produce the contour shown in Figure 11.1. Before the tests began, the firebrick had a thin slag glaze from previous development trials. The firebrick lining was inspected before the tests and again after the 67 hour main run. In general, the lining had the same appearance, with neither gain of slag nor loss of refractory, except for the 12 inch zone above the coke air ports.

Undoubtedly the hottest part of the coke bed was within 12 inches of the air ports and may well have reached 3000°F, with the refractory temperature several hundred degrees less. The slag was thinnest in the hottest zone and most conducive to penetrating and interacting with hot refractory. Fluorspar is also known to attack hot refractory by producing a lower melting slag with the refractory. Furthermore, when the coke air rate was varied quickly over a wide range or cut off entirely, the most rapid fluctuations in refractory surface temperatures occurred in the lower zone. The net result was an increase in lining diameter from 43 to 48 inches, as measured on two diameters 12 inches above the bottom of the coke air ports. This change occurred over three periods totaling 78 hours of combustion when the furnace charges totaled 155,000 pounds of refuse.

This rapid loss of refractory could not be tolerated in practice. However, it could probably be reduced or prevented by a redesign of the lower furnace zone to provide for a water cooled jacket with thin refractory lining that would achieve a thermal equilibrium with the coke bed and slag. Also, hopefully, the need for fluorspar could be reduced to rare occasions.

The temperatures measured at the refractory surface near the chute level, but 90 degrees away from the chute, reached 1400°F for a short time, but were generally only 900° to 1100°F during the best runs. However, temperatures of 1800° to 2000°F could be expected when the overall excess air rate is brought under control at 100 to 150 percent of stoichiometric for the total fuel and refuse charge.

Refractories can operate indefinitely at these and higher temperatures when the latter are reasonably steady, but slag attack from refuse constituents that melt on the refractory gradually destroys refractory. The service life of refractory above the coke bed can only be determined over periods of a year or longer.

ENERGY REQUIREMENTS

The electric power demands of the blowers, pumps and conveyors of the pilot plant were measured during the test. As all of the equipment were not in use all of the time, the power consumption was calculated for normal operation, assuming a power factor of 85 percent. The findings are listed in Table 7.

TABLE 7: POWER REQUIREMENTS

	Demand, Kw	Use, Kwhr
Coke air blower, 40 hp	17.70	17.70
Refuse air blower, 30 hp	18.35	18.35
Refuse conveyor, 10 hp	6.48	5.50
Coke elevator, 3/4 hp	0.75	0.13
Coke feeder, 1/3 hp	0.33	0.06
Frit conveyor, ½ hp	0.50	0.50
Frit tank pump, 5 hp	4.60	4.60
Cooling tower fan, ½ hp	0.50	0.50
Building, other, 120 amp	1.32	1.32
Maximum, total	50.53 Kw	48.66 Kwhr

It is not advisable to use these data to predict the power consumption for a full scale installation. Because of its experimental nature, the pilot installation was overpowered in some respects in proportion to tonnage throughput. Also, the pilot unit did not have a crane for moving refuse, nor was a dust collector in use. The power requirements of a full scale plant could be predicted from manufacturer's data on equipment to be used on a complete installation.

TECHNICAL EVALUATION

The technical advantages of the Melt-Zit process are a residue free of putrescible matter, maximum density in landfill, minimum pollution of ground water or streams, possible salvage or use value of the residue, and minimum use of water.

The capacity, refuse/coke/limestone ratios, and labor for Melt-Zit furnaces cannot be fully evaluated from the March 1968 tests. The incinerator would undoubtedly perform better with modifications, such as additional stir and feed controls, and others. Experience will also improve the skills of operators in managing the coke bed and furnace generally.

The methods used for conveying refuse, coke and residue will bear improvement. The top size of pieces of refuse must be limited to prevent clogging of the chute, or a method of mechanical feeding should be devised that will insure positive movement in the chute. The screw conveyor in an inclined trough jammed and had to be removed because all of the residue was not in the form of frit and small pellets.

A flight conveyor in the same trough did not fare any better. An earlier type of angle-iron drag conveyor which extended across the entire bottom and slope of the residue quench tank had the capability of moving coarse and fine residue without difficulty.

During the test the feeding of refuse was halted five minutes every thirty minutes while coke was being charged and heated. Simultaneous and steady feeding of coke, limestone, and refuse was tried but was not successful, probably because the furnace lining above the coke was not hot enough to ignite the refuse promptly. The method should be tried again after the present furnace has been improved.

High viscosity of glass and other high silica inerts is a fundamental problem of the Melt-Zit furnace. The prior removal of glass for salvage as saleable cullet, or separate collection as practiced in some localities, would be helpful. An increase in coke bed temperature through air preheat would be an economical method for reducing the slag viscosity. The addition of limestone should also be helpful in lowering slag viscosity, provided there is opportunity for the lime to mix with the glass. Higher temperatures are necessary for the lime to be fully effective. Preheating of the coke air would be helpful, and is essential if the coke rates are to be reduced to a minimum. Ample excess heat is available. Each 100°F rise in blast temperature would raise the coke bed, slag, metal and gas temperature by about 35°F. Again, only experience will tell whether the effect will be sufficiently beneficial or whether the viscous zone will only move higher in the coke bed. A modest air preheat would probably not be worthwhile, but coke air at 800°F would be effective.

A decrease in coke rate to 1/25 of the refuse is theoretically possible, and a coke rate of 1/20 of the refuse may be feasible in practice, even with a high moisture refuse. However, a decrease in coke consumption will be accompanied by a higher ratio of metal, inerts, and limestone in the coke bed, and lower temperatures in the coke bed. Air and gas friction during upward flow in the coke bed resist the downward flow of slag. These factors oppose successful operation.

The loss of refractory experienced could not be tolerated in a municipal plant. In order to gain sufficient refractory life a thermal equilibrium layer between fluid slag and solid refractory is required. A method that has been used for slag tap furnaces is forced cooling behind a relatively thin refractory.

Assuming that the above problems can be solved and demonstrated, one can estimate the capacities of similar furnaces of larger diameters. The upper limit in a circular furnace is a coke bed without a dead central core. A maximum practical diameter would probably be 10 feet, and possibly 8 or 9 feet. Furnaces of rectangular section are also possible.

A limit must be placed on the exit gas velocity, or the carryover of burning paper and fly ash will place an excessive burden on the secondary chamber and dust collector, as well as to seriously reduce the amount of inerts converted to slag. The test indicated that the primary furnace volumes required will be equal to those of conventional incinerators of the same capacities.

Excluding the metals, at least one-fourth of the mineral matter charged to the test furnace was blown out with the fly ash, with combustible in addition. The combustible could be largely burned out in a secondary combustion chamber of suitable size and design. The mineral matter would be caught, except for minor loss to the atmosphere, by the gas scrubber. The captured fly ash would have properties different from the slag, and would have to be reclaimed from the water.

Smoke from the Melt-Zit furnace may be expected at times as from the primary combustion chambers of conventional incinerators. A secondary combustion chamber will assist in smoke prevention.

If one assumes a maximum refuse rate of 4,000 pounds an hour for the test furnace of 3 feet throat diameter and 5 feet top diameter, the exit velocity will be close to 1,000 fpm. The refuse analysis is assumed the same as during the March 1968 tests. The gas volume will be proportional to the refuse burned, but the furnace areas and refuse tonnages will increase as the diameters squared. On this basis projected tonnages for various furnace sizes are given in Table 10.

TABLE 10: PROJECTED REFUSE TONNAGES AND COKE CONSUMPTION

Furnace Throat, inside dia., ft	Coke Bed dia., ft	Furnace top inside dia., ft	Projected Refuse Input, tons/hr	Coke at 1:20 refuse, lb/hr	Coke at 1:8* refuse, lb/hr
3	3.6	5.0	2.00	200	500
4	4.8	6.67	3.58	358	895
5	6.0	8.34	5.55	555	1388
6	7.2	10.0	8.00	800	2000
7	8.4	11.67	10.9	1090	2725
8	9.6	13.35	14.2	1420	3550

*Test B

The above hourly charging rate for refuse is 393 lb/ft^2 of coke bed which would be 25 percent better than was accomplished during Run B of the test. Improvements, tests, and studies suggested for the development of the Melt-Zit system include:

(1) Preheating of the coke air to 800°F or more by waste heat from the exhaust of the secondary combustion chamber. The objective is to increase the heat available above the 2600°F datum. Preheat of the refuse air is not recommended except as it may aid under some conditions; furnace exhaust gas temperature should not exceed 1800°F, as a deterrent to slag deposits above the coke bed.

(2) Refractory cooling around the coke bed to maintain the integrity of the refractory. Refractory compositions to be investigated for the critical zone.

(3) Improved and positive control by the furnace operator of the rate of feeding by weight of refuse, coke, and limestone to the furnace. Independent and prompt control of the rate of air supply to individual coke air ports, to the refuse air ports as a group, and to the furnace via the refuse chute. Independent control of furnace draft.

(4) Increased turbulence in the shaft furnace, above the refuse chute, to hasten combustion of entrained solids.

(5) Ample secondary combustion chamber for burnout of solids emitted from furnace. The adequacy of the existing secondary combustion volume was not determined.

(6) Reduction in coke consumption by reduction in coke air rate, coupled with preheat of coke air.

(7) Improved instrumentation of air flow rates and pressures, feeding rates, furnace radiation, radiation from coke air ports, refractory

temperatures, drafts, and solids carryover.

(8) The use of limestone to flux the molten glass and ash should be investigated to establish the conditions under which such use is beneficial or necessary.

(9) Commercial use of the slag and metal products should be explored to determine whether they are a source of potential revenue to offset the cost of coke. The effect of slow cooling on the strength of the slag would be of interest if the slag is to be used as an aggregate.

(10) The removal of fly ash from the scrubber water, and its disposal, will require attention, especially because of the high loss of mineral matter from the furnace.

CONCLUSIONS

In its state of development and method of operation during the March 1968 tests the pilot incinerator did not perform satisfactorily or reliably as evidenced by:

Low refuse firing rate. The best period of 6.5 hours had a refuse firing rate of 3,200 lb/hr (1.6 tons/hr). It may not be practical to exceed this rate by more than 25 to 30 percent without excessive entrainment of solids in the furnace gases.

High coke consumption relative to refuse incinerated. The most favorable rates were 8 tons of refuse per ton of coke during 6.5 hours and 6 tons of refuse per ton of coke for 11 hours. Too much air had been supplied to the coke bed.

Inability to drain molten iron and slag as necessary to maintain continuity of operation. The longest run was 62 hours during which 73 tons of refuse were charged. This run and others ended with refusal of the slag and metal to drain. The molten residue was too viscous. During periods of good slag flow the slag temperature by immersion thermocouple averaged 2532°F, with a peak of 2650°F.

Limestone in the amount of 1 ton for 38 tons of refuse was used for promoting slag fluidity. Seven pounds of fluorspar were also used per ton of refuse, in an attempt to gain complete slag control. Fluorspar additions should not normally be required.

The emission of fly ash, including combustibles and ash was 288 lb/hr or more, roughly equal to 10 percent of the refuse charged. Upwards of 25 percent of the refuse inerts may thus not be converted to slag, unless reintroduced into the furnace in a manner that would insure melting rather than reentrainment in the gases.

High labor requirement (3 men per shift) to manage the furnace proper. High consumption of oxygen and iron pipe lances to maintain slag flow.

The refractory furnace lining in the lower part of the coke bed was attacked by the high temperatures and slag ingredients. The lining in this critical area where the greatest refractory loss would occur, lost 2.5 inches on the furnace radius, as measured 12 inches above the

> lower air ports. The refractory in the upper part of the furnace had not deteriorated because the gas temperatures were seldom above 1900°F and usually below 1300°F during the test.

American Design & Development Corporation, the builders of the facility, however, did not agree with this conclusion by the consulting engineer for the project. Here are their comments concerning this conclusion of the consultant:

> He concludes the upper "refractory had not deteriorated because" soley due to temperature range experienced. We think that conclusion implies, first, that the refractory would deteriorate due to temperature alone and, of course, this would not necessarily be true, obviously the quality of material as well as temperature and, perhaps, even erosion would be factors.
>
> This form of conclusion also begs the question - what temperatures should be experienced at this elevation? In fact, the conclusion as stated involves consideration of a number of factors which cannot be accurately described purely from the test data. This sort of conclusion, we believe, could only be validly expressed after further study, tests and observations over an appreciable period of time under normal operating conditions.
>
> Continued tests of the Destructor (furnace) operation, including several runs in excess of 24 hours with sustained temperatures above 2000°F, since the March tests, have not resulted in observable or measurable loss of refractory in this area. We also think the reference to the upper refractory should be stated separately from consideration of the melt zone refractory. The conditions in the two areas are obviously quite different with different factors influencing refractory life.
>
> We strongly urge restatement of his last sentence and would suggest the following sentence be substituted as a more accurate and valid expression: "The refractory in the upper part of the furnace had not deteriorated."

The molten residue from the Melt-Zit unit was free of putrescible and combustible matter of negligible solubility, that could be deposited in landfill without cover. It would not attract vermin or rodents, and it was odorless and sterile. The slag became friable, black, glassy granulate when quenched in water. When cooled more slowly, it became a friable solid. It could also be flowed into dry molds and allowed to congeal and cool slowly in large masses.

A high yield of iron from the refuse was produced in the form of spherical and flattened drops, when the molten slag was run into water. About 20 percent of the melt weight was iron in this form, which could easily be separated magnetically from the granulate. The electric power consumption of the pilot plant was 48.66 kw/hr of operation. This quantity is not indicative of the power consumption to be expected in a full scale installation. The water in the residue quench tank was cooled by circulation through a blower cooled heat exchanger. Hence, furnace water consumption was the moisture on the granulate, about 14 gal/ton of refuse. The refuse burning rate is limited by the rate of melting of the inerts and/or by the loss of solids as carryover from the furnace. The carryover increases rapidly above 12 ft/sec of gas velocity as loose paper, foil, plastic film wrappings, sweepings do not fall to the coke bed but are immediately carried upward from the charging chute.

Based on the tests, the burning rates for several sizes of incinerators would be as follows, scaled up in proportion to nominal diameter squared:

Furnace Diameter, ft	Top Diameter, ft	Tons Refuse, per hour
3	5	1.60 (Test B)
4	6.7	2.84
5	8.3	4.45
6	10.0	6.40
7	11.7	8.72

When improved performance can be demonstrated, the rated tonnages could be increased, probably up to 30 percent. The tonnages could be increased until the amount of solids in the exhaust becomes intolerable. The tonnage capacity will decrease with lower refuse moisture content, as more air is required to burn the drier refuse and the entrainment of solids will increase. However, combustion will improve with drier refuse.

The Melt-Zit shaft furnace is only the primary combustion chamber of a complete incinerator plant. A secondary combustion chamber or chambers will be required to complete the combustion of the carryover solids. A dust collector will be required after the secondary chamber to reduce the particulate loading of the exhaust gases. Such equipment was in existence at Whitman at the time of the tests.

Restoration of Oil-Contaminated Beaches by Incineration

The material in this chapter is derived from PB 198,227 November 1970 prepared for the EPA by R.M. Roberts and T.S. Hoyt of the Envirogenics Company.

The cleanup of coastal areas contaminated by petroleum spills is as unpleasant as it is expensive. If not properly conducted, a renovation effort can result in devasting effects on the biota, serious depletion of recreational sand inventories, and lingering contamination problems that are not immediately obvious. The development of safe and more cost-effective techniques is being vigorously pursued. An objective is the development of capabilities wherein spills can be isolated and collected at sea before shore areas are befouled. However, some incidents of oil stranding will be unavoidable, as when the petroleum issues from a shore facility or beached vessel, or from an off-shore source close to land.

There are several programs dealing with new shore renovation concepts. A definite trend being established by this new technology is toward processes that avoid the sacrifice, by disposal, of valuable natural resources, notable among which is the sand of recreational beaches. In line with this objective, this is an investigation of combustive techniques.

Although any form of oil burn-off will certainly destroy organisms present, it is a process that would involve only a limited segment of the riparian zone. In any case, few totally nondestructive processes can be conceived of and these are slow. Relying on the washing action of the surf and/or biological effects to remove stranded oil can usually not be accepted if recreational areas are to be quickly returned to public use.

Depending on the nature of the problem, a variety of approaches are possible in considering the use of combustion processes for beach decontamination. Even when the proper modality is recognized, trade-offs still must be considered since the conceptual design inputs can be influenced by many economic, thermodynamic, and operational constraints.

SYSTEM DESIGN

Overview

Initial tests proved that oil-contaminated sand can be thermally cleaned to furnish a product that is hardly distinguishable from sand that has not been contaminated. This testing also provided data needed for the design of practical incineration systems; this included information pertinent to handling times, fuel and power demand, heat penetration, and temperature requirements.

These tests further established that the contaminated sand progresses through three stages in the thermal cleaning process. The first stage involves dehydration and oil dispersion. It takes place when the oily sand requires a temperature of about 300°F, wherein the oil-fraction becomes quite fluid and coats the sand and excess oil flows from the mixture. When the temperature of the system is increased to about 600°F, all of the oil fraction will vaporize and/or pyrolyze to leave a black solid coating on the sand grains.

It was established that if the input material (moisture content $\leqslant$6 wt %) is surrounded by hot sand or air at 700°F, the first two effects will be accomplished in approximately two minutes. This parameter, however, is a function of heat transfer rate, which in turn is affected by feed rate, apparatus configuration, air flow, feed material characteristics, and many other variables.

The vapors, other than water, emitted during the first two stages of transition have considerable fuel value; utilization of this energy source is of prime importance in apparatus design.

The final cleaning stage consists of exposing the black, dry sand to sufficiently high temperatures and sufficient air to burn off the carbon coating the sand grains. Tests showed that this could be accomplished in about 60 seconds at 1400°F. Here again, many variables will influence the time requirement.

In all of the stages of cleaning it was found that agitation of the oil/sand mixture was required in order to avoid the formation of undesirable clinkers and to obtain reasonable cleaning rates. The known methods of accomplishing the required heating and agitation considered on this project were: rotary kiln, fluidized bed, and cyclonic cleaning.

The input material to be processed must be expected to vary considerably in its chemical and physical properties. In terms of mechanical state, the input may vary from a fairly fluid solid/liquid mixture to a feed which contains large plates of rigid asphaltic contaminant. The input should normally contain more than sufficient ($\geqslant$6 wt %) petroleum to permit the Stage 2 transition to occur without introducing supplementary energy. The moisture content of the input is a critical parameter and should be held to a minimum. It is not unreasonable to expect, however, that an average of 6 wt % moisture can be obtained if the excavation operation is managed in a specific manner.

Another important characteristic of the feed material will be manifested as unavoidable inclusions of variously dimensioned foreign and natural objects. With the obvious exception of gross pieces, the rejection of such items prior to combustive processing would not be possible. The incinerator would therefore have to be designed to screen the process material at some point between input and output, or to permit the complete passage of all noncombustibles that are introduced.

Finally, throughput requirements must be considered. Unfortunately, the linear rate of beach cleaning cannot be predicted on the basis of the quantity of sand that can be decontaminated per unit time. A number of factors preclude such a correlation. It is nonetheless pointed out that under conditions that one can reasonably expect to find, a feed rate of between 38 and 53 tons per hour will result in the decontamination of 1 mile of beach per 8 hour work day. Assuming that this linear renovation rate is an acceptable one, the input specification for a full-scale system has arbitrarily been set at 60 tons/hour.

Design Analysis

Stage 1 and 2 transitions of the cleaning process are the most difficult to accomplish in a short time, primarily because of configurational constraints. The rotary kiln represents, by far, the readiest solution to these problems; it provides containment of the materials, while allowing agitation, conveyance, and gas/vapor containment. For this reason, process design was developed on the basis of the use of a rotary kiln type of apparatus for at least the production of Stage 1 and 2 material.

Fluidized bed methods are feasible for the final (Stage 3) cleaning. It was calculated, for example, that only 15,000 ACFM of air (1400°F) would be required to fluidize a one-ton sand bed. Assumed here is an average carbon-coated particle diameter of 0.03 inch, a solid phase density of 100 lb/cu ft, and a superficial gas velocity of 4.5 ft/sec. This volumetric flow rate (and, thus, the fan power required) is considerably lower than would be necessary for a rotary kiln configuration of the same output capacity.

A disqualifying drawback of the fluidized process is the fact that the input would have to be screened so that the bottom of the bed would not accumulate a blockage of larger-diameter noncombustible material. The utilization of a screening process was studied in some detail but it was concluded that the incorporation of such components would create more problems and functional risks than the use of a fluidized bed could justify. Specifically, no suitable arrangement could be envisioned that would permit the simultaneous screening and rejection of oversized objects and the collection of decarbonized sand elutriated from the bed.

Cyclone Cleaning

While considered in the experimental portion of this project, cyclone cleaning was found to be practical only for combusting Stage 3 sand. Because of the extreme abrasion problems encountered in handling sand in a cyclone cleaner, and in view of the fact that the initial cleaning stages would still have to be accomplished in a rotary kiln type of device, this method was not considered further. This analysis therefore considered design parameters and configurations based exclusively on the employment of a rotary-kiln type of thermal cleaning device.

Batch Kiln Process

This process would consist of the use of a single rotary kiln, in which an amount of contaminated sand would be loaded, processed and the clean sand periodically dumped. While production rate would be low, this type of process does have certain advantages. For example, control problems are negligible and material handling would be simplified. This type of processing could be accomplished by modifying presently available equipment, specifically a cement mixing truck. By adding a refractor lining, burner, fan, and a fuel atomizing pump and compressor, it would become a batch-type rotary kiln. If air pollution problems are to be considered, an afterburner would definitely also be required. Figure 12.1 illustrates how these modifications could be accomplished.

FIGURE 12.1: BATCH TYPE SAND CLEANER

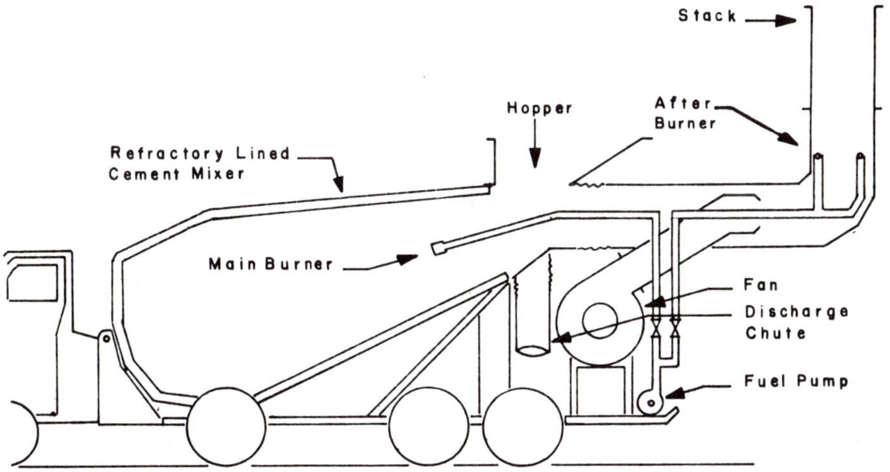

Considering a 10 cubic yard transit mixer, at 20% fill, it is estimated that 2,800 lb/hr of sand could be processed in a single mixer. Thus it would require 7 mixers this size to produce an output of 10 ton/hour. Fuel requirements for this processing rate would definitely be higher than for the continuous process (discussed in the next subsections), since very little of the heat available in the oil contaminated sand would be utilized. Power consumption, however, would be somewhat lower. Labor costs would obviously be higher. Only a single operator will be required for a continuous system (sub- or full-scale), while at least two men would probably be needed to tend every four or six trucks.

Equipment costs can be estimated based on two approaches: the purchase and modification of new equipment or the temporary retrofit of rented vehicles. If the former practice were followed, the equipment cost would be considerably higher than for a continuous system. The latter (rental) approach would be far cheaper but would entail problems, e.g., delay while retrofitting and acquiring suitable vehicles.

CONTINUOUS KILN OPERATION

Process Description

Rotary kilns, both refractory-lined and unlined, are common items of industrial equipment. The kiln consists of a tube rotating on its longitudinal axis, through which solids or slurries are conveyed with or against a stream of process gas. The interior of the shell can be equipped with a variety of material lifters, baffles, conveyors, or material showering arrangements to provide virtually any desired material flow characteristics.

In general, the function of the rotary kiln is insensitive to the material being handled; rocks, sand, bottles, seaweed and driftwood (within reasonable limits) will not affect operation adversely. In this process context, both countercurrent and concurrent gas/feed flows are applicable. The concurrent arrangement is indicated for the first portion of the system (stages 1 and 2) in order to remove volatilized oil constituents and pyrolyzates from the sand without recooling, while countercurrent flow is suggested in the high temperature stage, in order to operate at lower gas temperatures. In addition to this two-part rotary kiln, a combustion chamber is needed to provide heat for the process and to burn oil vapors given off in the pyrolysis stage.

Figure 12.2 shows the basic flow in this type of apparatus. The relative flow rates from the stack and rotating-cleaner blower would be determined by the speed of the latter, which, in turn, would be controlled by gas temperature. Normally, the feed for this process from a hopper or conveyor would be continuous. An apparatus producing the indicated effects would be sufficient to clean sand. To minimize fuel requirements, however, several alternatives are available which should be considered for the final design. These are: (1) Hot sand recycle, (2) Hot air recycle, (3) Fresh air preheat with sand cooler. All of these alternatives may be practiced on a continuous feed basis. Information, data, drawings, etc., discussed herein are based on a nominal (minimum) clean sand output of 10 tons/hour, rather than that of a full-scale system. This was done because the output estimates are highly conservative and actual results obtainable could well exceed (by as much as a factor of three) performance expectations. In addition to avoiding oversizing, it was the intention to provide characteristics for a system suitable for early field test evaluation which would involve a more reasonable R&D capitalization.

FIGURE 12.2: BASIC CONTINUOUS SAND CLEANING FLOW DIAGRAM

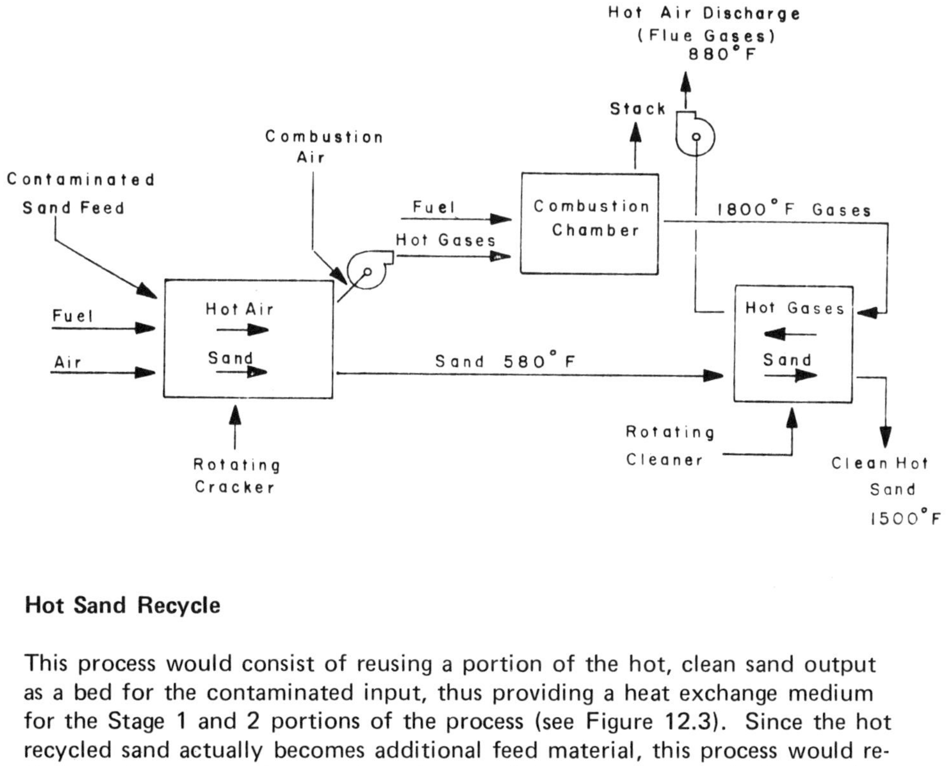

Hot Sand Recycle

This process would consist of reusing a portion of the hot, clean sand output as a bed for the contaminated input, thus providing a heat exchange medium for the Stage 1 and 2 portions of the process (see Figure 12.3). Since the hot recycled sand actually becomes additional feed material, this process would reduce the input rate of contaminated material by an amount equivalent to the amount of sand recycled.

The volume of the kiln would therefore have to be enlarged by a factor representing the square root of the fractional throughput displaced by recycle sand. The amount of recycle sand needed will depend on the heat required to produce the first two stages of sand renovation. If no auxiliary burner is included to produce the first stage transition, this would almost equal the input of

FIGURE 12.3: HOT SAND RECYCLE FLOW DIAGRAM

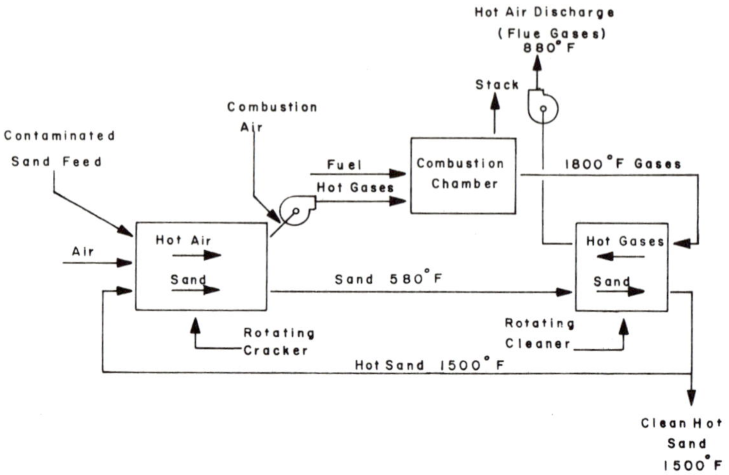

contaminated sand. Because of the equipment size limitations imposed by the mode of application envisioned, this concept has been reserved but not discarded insofar as the level of design effort is concerned.

Hot Air Recycle

This effect would consist of utilizing a portion of the output from an augmented hot gas generator to provide heat, not only for the final sand cleaning stage but for the initial stages of cleaning. This arrangement would eliminate the requirement for an additional burner and make the use of hot sand recycle unnecessary. Figure 12.4 diagrams the concept.

FIGURE 12.4: HOT AIR RECYCLE FLOW DIAGRAM

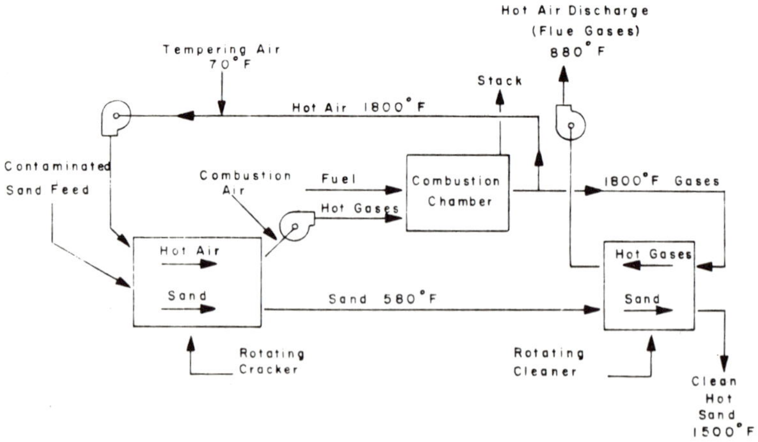

The preheat air temperature can be raised to any level the physical limitation of the equipment will allow and would be maintained constant by the addition of tempering (fresh) air ahead of the fan. This arrangement would necessitate an increase in the size of the equipment, so that the linear rate of gas flow could be held below the conveying velocities. For this purpose, 750 ft/min was selected as the maximum allowable velocity through the kiln.

Fresh Air Preheat with Sand Cooler

If the clean sand were discharged at final process temperature (about 1500°F), certain handling problems and safety hazards would exist. This difficulty would be considerably reduced by the addition of another rotary kiln, through which clean fresh air were passed countercurrently against the discharging sand flow. Sand effluent temperatures as low as 250°F could be reasonably obtained in this manner. The exhaust air, which would be heated to about 1000°F in the process, could then either be discarded or, to conserve fuel, used in the first stages of the cleaning process.

Equipment Description

A complete sand cleaner, as herein envisioned, consists of: cracker, combustion chamber, cleaner, appurtenant equipment (fans, burners, fuel source, power supply and controls) and an optional sand cooler. A brief description of each follows.

Cracker: This is a rotary shell wherein the first two stages of cleaning take place. Contaminated sand is fed into one end through a vertical hopper, along with sufficient concurrent hot gas to complete Stages 1 and 2. The heat is supplied either by a burner, hot recycled sand, or hot recycled air. Approximately 6.2×10^6 Btu/hr would be required. This figure includes no credit for oil combustion energy and assumes an input of sand of 20,000 lb/hr (inert basis), together with the maximum expectable amount of oil (5,000 lb/hr or 20 wt %, dry basis) to be heated (but not burned) and water (1600 lb/hr or 6 wt %) to be vaporized and heated in both the liquid and gaseous phases. This would, of course, be decreased by any combustion heat that is provided by the oil and/or reduction in water content.

Since, at the probable minimum oil content (6 wt %, dry basis), the oil will release about 23×10^6 Btu/hr in the presence of sufficient oxygen, it will be necessary to limit the amount of air that is introduced to prevent complete combustion from taking place in this stage. If the full 6.2×10^6 Btu/hr required in this stage could be obtained from the oil combustion, it would require (at 1 cu ft air/100 Btu) 1,030 SCFM air in this zone.

Similarly, if no oil were present, an auxiliary heater system would be required to furnish a like amount of air. The off-gases from the vaporizing and pyrolyzing oil (maximum amount) taken together with the combustion air and water vapor produced would represent a maximum volumetric flow, at 600°F, of about 11,000 ACFM. This would have to be ducted in a kiln diameter of no less than 4.5 ft in order to maintain the gas velocity through the shell at less than 750 ft/min. Length, based on 10% fill (normal for rotary kilns) and a retention time of 6 minutes (furnishing a 3x safety factor) would be about 15 feet. Since the temperatures in this portion would probably not exceed 800°F, the shell could be constructed of mild steel and without incorporating a liner. Lifters would be provided for proper agitation and transport.

Combustion Chamber: The combustion chamber for furnishing process heat would be a nonrotating, refractory lined cylinder, sized to accommodate a heat release from combusting petroleum vapors as high as 9×10^7 Btu/hr, based on a maximum envisioned oil input of 5,000 lb/hr. An auxiliary burner would be included to provide heat for the process when the oil content of the sand is low. This combustion chamber would operate at sufficiently high temperatures to avoid air pollution problems. The discharge gas would be maintained at about 1800°F by the addition of appropriate quantities of diluent air. A spark ignited continuous pilot would be provided to insure a source of ignition. Combustion air for the final cleaning stage would be supplied by a fan, which would also remove vapors from the cracking stage.

Sized for maximum required heat release, the combustion chamber would have a volume of about 530 cu ft. It would be a 20 ft long cylinder, 6 ft in diameter, lined with a 4.5 inch thickness of refractory material.

Cleaner: The purpose of the "cleaner" is to accept the black, carbon-coated sand flowing from the cracker and increase its temperature from about 600° to 1500°F, thereby oxidizing the carbon from the sand grains and furnishing a discharge of clean 1500°F sand. Hot (1800°F) air from the combustion chamber would flow countercurrently with respect to the sand flow in this kiln. Because of the higher operating temperatures, the rotating shell for this stage should be constructed of a high temperature alloy.

This design feature would probably promote minimum discoloration of the output. Dimensioning of the shell would be on the basis of a 10% fill, 3 minute retention time, 750 ft/min gas velocity, and a temperature drop of 900°F in the hot gases as heat is acquired by the sand. About 4×10^6 Btu/hr must be transferred to the sand; this would require about 4000 SCFM of gas, or 18,000 ACFM at 1800°F. To maintain a gas velocity of 750 ft/min in the shell would require an inside diameter of 6 ft. Because, however, the gases are cooled in passing through the kiln, the volumetric discharge of gas would only be 10,000 ACFM (at 900°F).

Thus, if the cleaner were sized down to have, conveniently, the same diameter (4.5 ft) as the cracker, the discharged gas velocity would be 630 ft/min. This would preclude the blow-back of any solids into the cracker. At the (1800°F) sand discharge end of the kiln, however, conveying velocities would be exceeded. This would result in some material being carried back into the cleaner, but not its full length. A zone containing a heavier concentration of sand fines would be formed but the normal output flow would be reestablished after steady state conditions were developed. The 4.5 ft (diameter) shell would require a length of 7 ft, in order to allow room for drive mechanisms, trunnions, etc. Rotary drier experts advised that a length of 12 feet would be more suitable for this shell. Lifters would be provided in this shell to obtain agitation. External shell temperatures could be reduced with refractory, if required.

Appurtenant Equipment: In addition to the three components previously discussed, several major items of equipment would be required to complete the system. These would include: air handling equipment, burners and fuel source, power supply and controls.

Air Handling Equipment: Fans will be required to provide hot air for the cracker and the combustion chamber, and motive power for air flow in the cleaner. Two fans will be required to perform these functions. One fan would supply combustion air (15,000 SCFM) to the combustion chamber and draw air and vapors through the cracker. The other fan would draw air through the cleaner (4,000 SCFM). Both should be of high temperature alloy construction for maximum service lift and to permit the handling of the high temperature air. The first fan would require a 15 hp motor, and the second a 5 hp motor.

Burner and Fuel Source: Because of its availability, diesel oil has been selected as the auxiliary energy source for the machine. Air-atomized oil burners would be used to provide heat injection. Two burners would be required in the basic unit; one in the cracker and the other in the combustion chamber. The maximum heat required for these stages, when none is available in the sand influent, would be slightly over 10^7 Btu/hr, or 72 gph of oil. Diesel fuel would also be used to supply a motor/generator set to furnish electrical power to the fans, pump, compressor, kiln rotating motor, and controls. A hydraulic pump and motors might also be used for this purpose. A total of 45 hp would be sufficient to provide the required power, which is itemized as follows:

Component	Horsepower
Fans	20
Pump	5
Compressor	5
Kiln Drive	10
Control and Miscellaneous	5
Total	45

The fuel consumption of the motor-generator set would be approximately 2 gph.

Controls: Two temperature control systems will be required. One would control the temperature of the cracker by varying the fuel and air fed to the burner. The second would control the hot sand discharge temperature of the cleaner by varying the temperature of the gas led into the cleaner. This would be done by adjusting the amount of dilution air mixed with the gas discharged from the combustion chamber. In addition, temperature limited fail-safe devices should be provided in each chamber to prevent equipment damage in case of malfunction. Each burner should also be equipped with combustion safeguards to insure safe operation.

Sand Cooler: The physical size of the optional sand cooler would depend on the maximum allowable sand discharge temperature. If the cooler were the same length and diameter as the cracker and cleaner together, retention time would be 9 minutes. If the air is drawn through the cooler (as shown in Figure 12.5), the maximum air temperature rise permissible through the cooler would depend on the limitations of the fan. Assuming a maximum fan temperature rating of 1000°F and an input (hot day) air temperature of 100°F, the ΔT would be 900°F. If the exiting gas velocity were controlled at 750 ft/min, the volumetric flow would be 12,000 ACFM at 1000°F, or 4,250 SCFM.

FIGURE 12.5: FRESH AIR REHEAT WITH SAND COOLER

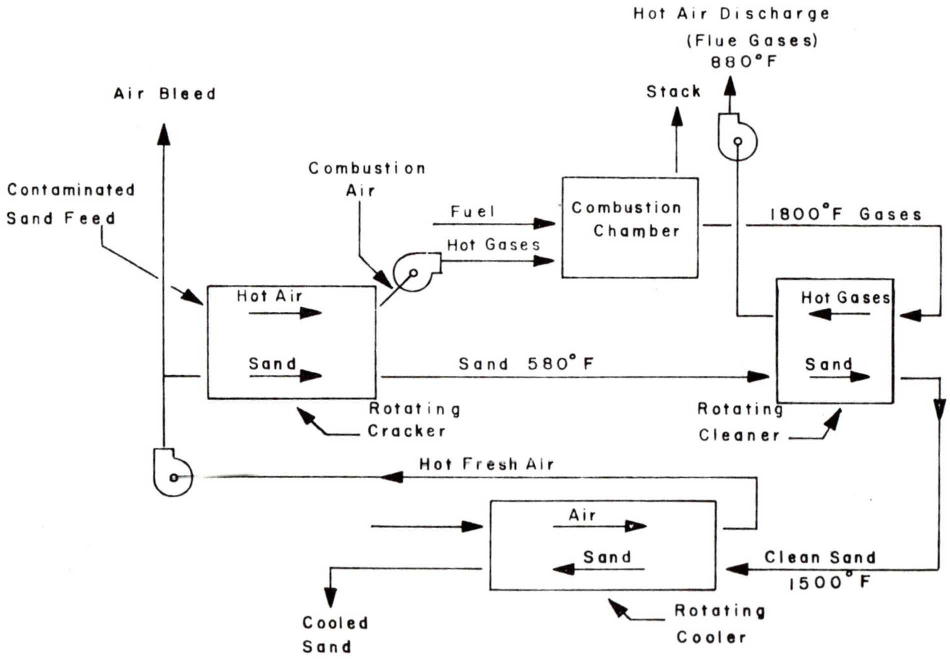

Based on the given ΔT, the heat transfer to the gas phase (neglecting small heat capacity corrections for changes in chemical composition) would be about 4.3×10^6 Btu/hr. This would be sufficient to lower the sand temperature from 1500° to about 415°F. This could be further reduced by increasing the diameter of the shell and (correspondingly, based on cross-sectional air increase) the air throughput.

Another approach to achieving cooler sand output would be to relocate the fan from the gas outlet to the air input. This would allow the outlet gas to come to much higher temperatures. It might also pose design problems if the system proves to have a high Reynolds number and considerable air is lost through the sand discharge opening. As previously mentioned, however, sand temperatures as low as 250°F are not unreasonable to expect. If, in fact, this parameter becomes important in terms of operational efficiency or hazard control, a simple remedy would be to spray sea water onto the piling output. This could be done with minimum increase in operating cost. Such a process might also be worthwhile to minimize thermal damage to the output conveyor belt.

System Configuration

Figures 12.6 and 12.7 illustrate two possible arrangements of the apparatus and indicate approximate sizes. Figure 12.6 shows the basic unit, in which heat recuperation from the output sand is not involved. Figure 12.7 indicates the size increase required for the addition of hot sand recycle. Addition of hot air recycle is not shown.

However, the size would be approximately that shown in Figure 12.7. The system size will increase approximately as the square root of throughput, so that 100 tph units will be approximately three times as large as the 10 tph systems illustrated in Figures 12.6 and 12.7. The weights of these combustors have also been estimated and are shown as a function of throughput in Figure 12.8. As can be seen, a 10 tph system would weight about 13 tons and a 60 tph combustor about 38 tons.

Air Pollution Considerations

The bench tests conducted demonstrated that, if sufficient oxygen is provided for combustion, the oil in the influent can be burned with no visible particulate emissions. Dust generation from air flow through the sand is possible. In this case a small cyclone dust separator might be required. Fan B (see Figure 12.6) motor horsepower would then have to be increased to 10 hp to provide additional static pressure for flow through the cyclone.

Other Characteristics

It was felt that the detailed consideration of certain system characteristics would be premature at this time. Engineering aspects dealing with optimum transportability, beach mobility, and sand handling to and from the system were

FIGURE 12.6: BASIC ROTARY KILN SAND CLEANER

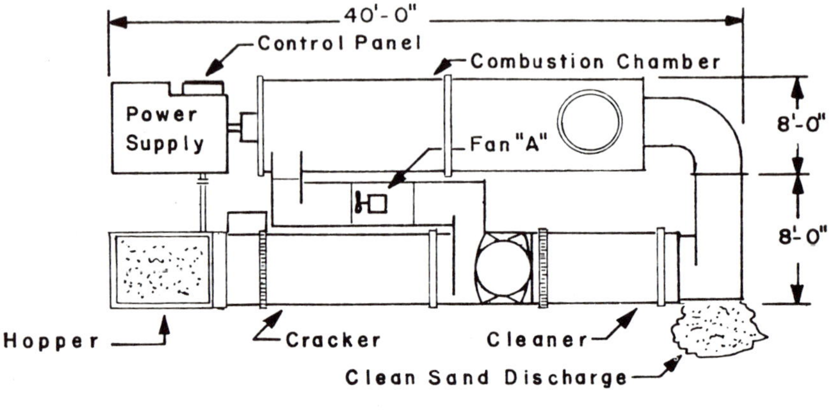

Plan View

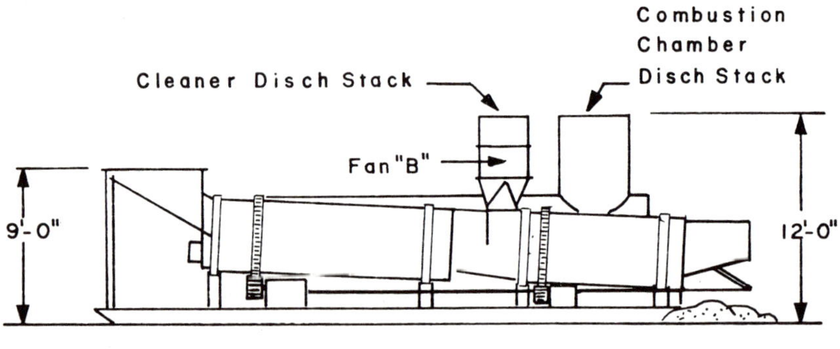

Side Elevation

FIGURE 12.7: ROTARY KILN CLEANER WITH HOT SAND RECYCLE

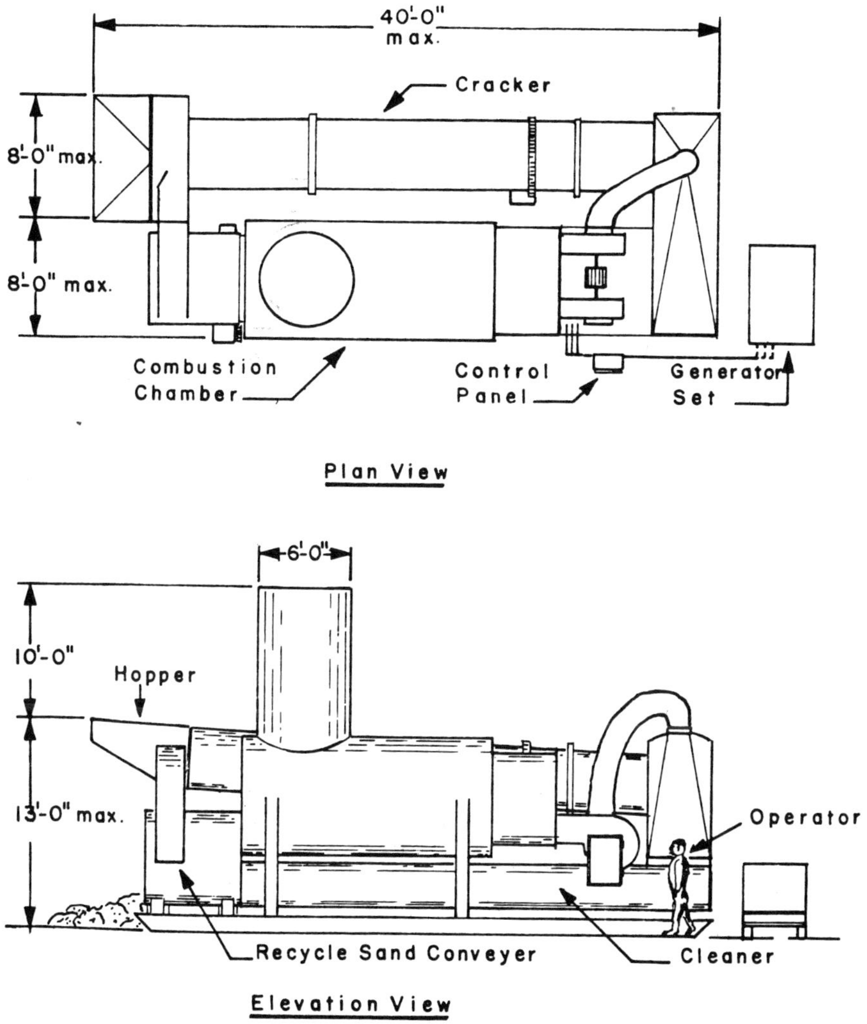

FIGURE 12.8: ESTIMATED WEIGHT OF BASIC SAND COMBUSTOR

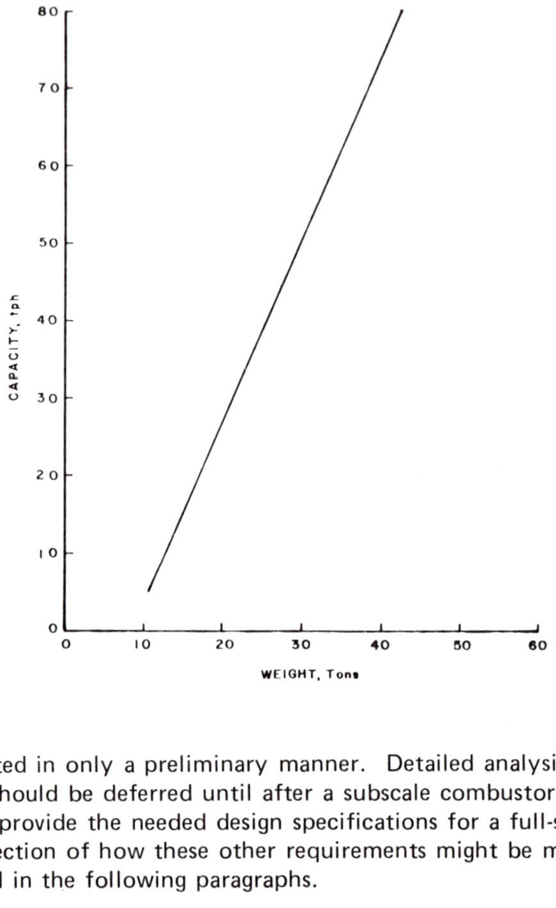

therefore treated in only a preliminary manner. Detailed analysis of these design features should be deferred until after a subscale combustor has been built and tested to provide the needed design specifications for a full-scale system. A tentative projection of how these other requirements might be managed is, however, discussed in the following paragraphs.

Transportability: Assuming that the subscale system is capable of operating at only its minimum rated throughput (10 tons/hour), a full-scale incinerator (38-53 tons/hour) would, theoretically, have to be 1.95 to 2.30 times larger. The dimensions of the subscale sand cleaner are, however, already such that transporting it by air would only be possible if the system were dismantled and moved in several aircraft. If still larger components are indicated, it is doubtful if air transportation would be physically possible or economically tolerable.

It may eventuate that the subscale system will satisfactorily perform at its maximum projected throughput (30 tons/hour) and that this production rate would be accepted as adequate for routine beach cleaning operations. Even if scale-up proved unnecessary, the question of whether flying such heavy and bulky equipment would be advisable must still be considered. In all likelihood, the sand cooler (if used), the combustion chamber stack, the pump and generator set, and, possibly, the cracker would have to have detachable features to permit separate or repositioned shipment. Providing this capability were feasible, it could increase equipment costs by as much as 25%. A further drawback would be that a significant portion of the time gained by air freighting the equipment would be lost in having to reassemble it at or near the use point.

Although trucking the intact equipment to a disaster scene might require an additional day or two, the device could be put onto the beach directly from the flat-bed carrier. The money saved per trip by this method of drayage could be considerable. Thus, a careful analysis of these factors will have to be made prior to hardening the ultimate configuration of the renovator and determining its mode of deployment. Another possibility is the shipment of the combustor from a seaport storage-point to a spill site by amphibious seacraft. The danger implicit in this kind of delivery is well known even for specially-designed (combat) vehicles. A tractor-drawn, nonruggedized machine might not survive an unfavorable landing situation.

Beach Mobility: The conceptual designs developed herein present an incinerator that is sled mounted. Such an arrangement will doubtless be adequate for restricted field test purposes. In actual renovation operations, however, the system will probably have to be moved several times a day in order to keep the incinerator reasonably close to the excavation equipment. While it would clearly be impractical to develop a self-propelled machine, it may be possible, under certain beach conditions, to operate the renovator on the bed of the carrier which delivers it.

However, additional provisions would still be needed to permit its easy movement along a beach by being drawn by a tractor or similar vehicle. A sled base may well prove inadequate for this purpose. An appealing traction form would be the tracked (continuous belt) assembly.

It should be pointed out that, in moving the machine along the contaminated area, it should not be allowed to cool. This would cause reheating delays and additional fuel costs. Thus, while being moved, the combustor would be fired by its auxiliary fuel source.

Sand Handling: The subscale system discussed herein is a hopper-fed device which discharges its cleaned output directly onto the ground. For field testing purposes it will be necessary to build up a sand ramp adjacent to the incinerator so that front end loaders can elevate their shovels above the 13 ft high hopper. Similarly, the sand discharged will have to be removed periodically to prevent the incinerator output opening from becoming blocked. Some type of sand-conveyance device will have to be provided when a fully operated system is designed. The configuration this ancillary equipment should assume will be better recognized after ultimate incinerator requirements are defined. It is expected, however that certain of the sand moving techniques being evaluated will be of value for that purpose.

Costs

The costs of fabricating test devices of the type herein described will be very close to those of fully functional units, excepting two cost factors. These are the feeding and outloading systems, and the traction assemblies, which have not been considered. Neither of these cost items should be excessive, but further design analysis will be required to verify this conclusion. The cost function for the basic system is shown in Figure 12.9.

In normal operating costing, many factors are applied that would be difficult to consider in terms of this system. The equipment described here would be employed for a relatively short time following an oil spill disaster and costs would be indemnified by the party found responsible for the damage. The responsible party could not be expected to pay for the equipment on a fair rate-of-return basis nor could he be allowed to share pro rata in capital cost annualization. The situation is somewhat analogous to that wherein the owner of a fire-damaged home would be required to pay the fire department for services rendered on an annualized basis.

FIGURE 12.9: ESTIMATED CONSTRUCTION COSTS OF BASIC SAND COMBUSTOR (1970)

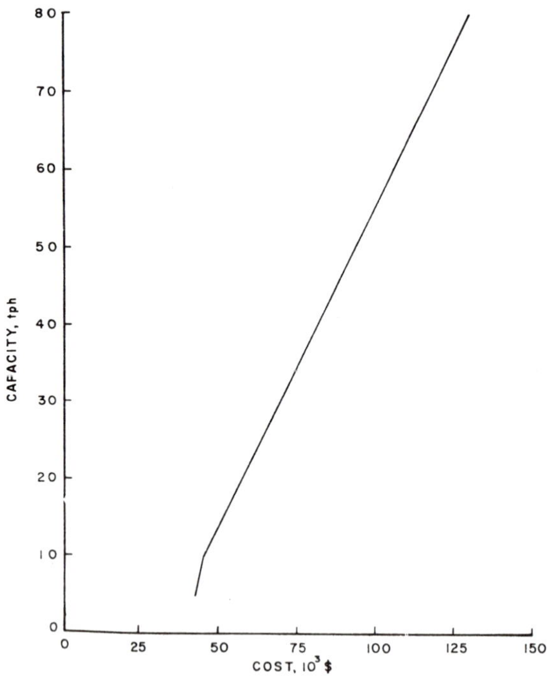

Annual capital costs are determined by annualizing the initial equipment investment, using an appropriate capital recovery factor and then adding to this rate due compensation for interest, insurance, taxes, etc. When these costs are returned by users, the equipment use-rate determines the degree of recovery that must be applied. If the equipment is used one month out of the year, the normal rental rate must be increased by a factor of 12.

Based on an estimated equipment life expectancy (N) of ten years, the amortization rate (A) would be 14.9%, if an 8% rate of return (r) is expected. That is,

$$A = \frac{r\,(1 + r)^N}{(1 + r)^N - 1}$$

Federal taxes should be set at about 3%, insurance at a minimum of 0.25%, and state and local taxes at about 1.9%. Annualization would then be 20%. From Figure 12.9 it is seen that a 10 tph combustor will cost about $45,000 in its basic configuration, and probably about $60,000 when equipped with a suitable feed conveyance system and an appropriate traction assembly, neglecting engineering and design costs. Annualization would therefore be $1,000 per month assuming 100% utilization (plant factor). If the equipment were only used on spill cleanups 30 working days per year, which is perhaps conservative, the annualization charge would be $5.00 per ton of sand, assuming an 8 hour work day.

If alternate schemes were developed to insure year-around use of the combustor, the annualization cost would drop to about $0.42/ton. This would require of course the utilization of the combustor in other practical process operations, as discussed later, when not needed for its primary purpose. Such an arrangement would be entirely feasible considering the function of the combustor, but probably impractical if the process in which it were alternatively used had to be shut down because of its sudden withdrawal for oil-spill emergencies.

Another aspect of the cost analysis that is difficult to quantify involves the loading and off-loading operations. Definite capital costs will be associated with the constant-feed input conveyor and the output system that will permit the cleaned sand to be disgorged at a reasonable distance away from the combustor. Design of such machinery was not attempted on the feasibility study.

Operating costs will be involved in the loading of the feed conveyor hopper and removal of clean sand deposited by the output conveyor. No additional costs can be identified with this operation. At a typical beach restoration site, front-end loaders would normally be deployed to remove contaminated sand piled up by graders. This material would then be deposited in trucks for outhaul or, in this case, at the working site of the combustor.

In the latter situation, the combustor would be fed directly by the front-end loader, at no cost penalty, if the load it had acquired had been resting on the berm for an hour or more. The renovation could be managed in such a manner that this operation could be performed without incurring additional costs. Returning cleaned sand to the surf would be done by the same frontend loader(s) that furnish the input to the combustor. A problem would be that the shovels would probably be filthy and tend to recontaminate the renovated material. This could be prevented by applying a suitable coating to the shovels.

POSSIBLE SECONDARY USES OF THE COMBUSTOR

The hardware described herein would hopefully have very limited opportunities of application. Like an ambulance, however, the annualization of the capital investment for the owner would have to be passed off as an unfortunate penalty to the sometime buyer of the equipment's service. This would predictably occur between periods of considerable inactivity. Secondary uses of the combustor thus would seem desirable to level this burden, but for the same reason that ambulances are not used to deliver groceries when idle, alternate utilization perhaps should be carefully considered.

Another constraint is that a combustive beach renovator would necessarily be a low capacity machine in order that it be deployable. It would therefore be of little interest to large-scale operators dealing with, say, ore-roasting, calcining, or process-slurry dehydration. In fact, it would doubtless be uneconomical to use in its primary function on beaches that had only experienced minor oil-spotting from the bilges of passing ships or from small, chronic submarine seeps.

Considered as a mobile incinerator, however, a number of specialty uses can be suggested that might be compatible, particularly on an interruptible basis. These include the following operations: Destruction of classified material; incineration of hospital wastes; neutralization of low-energy, toxic chemical/biological agents (military, agricultural, etc.); disposal of contraband (narcotics and other smuggled goods); rare-ore treatment; chemical waste disposal at special sites; PHS destruction of infected articles in plague areas; and on-site cleanup of littered areas following heavily attended mass meetings of people.

CONCLUSIONS OF FEASIBILITY ANALYSIS

Process Feasibility: The renovation of oil-contaminated shorelines using a combustive process is practical only under certain conditions. In-place cleaning with flames is not feasible; the contaminated beach material must be excavated and fed into the combustor. The sand cleaned by combustion does not entirely recover its original color; a slightly greyish tone persists due to tiny pockets and platelets of carbon which adhere to granules that are not of quartz origin. Abrasion in the surf would probably remove this source of slight discoloration.

System Configuration: The system recommended is a three-chamber combustor operating on the rotary kiln principle. In the first chamber, combustion air, recuperatively heated, would be introduced in concurrent flow with the oily sand. The feed would dehydrate, the oil become evenly distributed on the granules, and partial volatilization and/or pyrolysis of the oil fraction would occur. Before exiting the first chamber, the sand feed would achieve a carbonized condition and good flow properties. The resulting granular, carbon-coated sand would then be passed into a higher temperature kiln where the gas/solid flow would be countercurrent. Here the material would be decarbonized and cleaned to output condition. Heat could be extracted from the cleaned sand prior to discharge in a third, rotating section.

An output of 60 tons/hour is envisioned for a full-scale system. This should permit, with proper earth-moving practice, the renovation of a 20 ft wide strip of beach to a depth of 2" at a rate of 1 mile per each 8 hr work day. Based on conceptual drawings a subscale system (10 to 30 tons/hour) suitable for immediate field testing would be 40' long, 16' wide, and 13' high (excluding stack); it would weight 13 tons. Because of the large size of the system, self-propulsion would not be practical. The recommended method would be to position the combustor on a level portion of the beach and have front-end loaders bring material to it. As the cleanup operation progressed, the combustor would be periodically moved by a bulldozer. The input and output material would be handled by conveyor equipment. In terms of sand cleaning, the performance of the envisioned system should not be affected when drift trash and other beach debris is fed into it. This material should also emerge oil-free, but not all of it would be fully combusted. Such charred articles should be screened from the discharge.

Characteristics of the Feed: Only limited data could be obtained on the composition of sand removed from oil-contaminated beaches. These resulted from laboratory tests performed by Government technicians in connection with the cleanup following the Ocean Eagle disaster at San Juan, Puerto Rico. The many samples taken by these workers were described as resembling the material that was excavated for disposal. An average of 16 wt % of petroleum residues was observed. Estimates at another spill site ranged even higher, while others have suggested that the 16 wt % value is unrealistically high.

For combustion purposes, an oil content of 6 wt % is adequate for self-sustained combustion provided the moisture content is also about 6 wt %. Typical beach sand removed from the swash zone, and from which excess seawater has been poured off, will have a moisture content of 11 wt %. Piled on dry sand, such wet material will drain by capillary action to 6 wt % in 1 hour.

Equipment Deployment and Operation: Because of the envisioned size of the combustor, transportation to the spill site should be handled by firms specializing in the drayage of heavy equipment. It is doubtful that transportation by air would be practical. The width of even a 10 tph unit is such (13') that rail-haul would not be possible. The system could be designed to be dismantled for air- or rail-haul, but this increases capital costs by at least 25%. Truck drayage appears to be the only suitable means of cross-country deployment and this would probably require removal of the stack and the issuance of a special permit in most states because of equipment width. In moving the combustor between coastal points, serious consideration should be given to the use of barges. An actual landing of the equipment by amphibious craft at the beach site would also be possible, but only if the shore terrain were favorable.

The combustor would best find application on sand beaches of the recreational type which are reasonably accessible by good roads. The beach itself should have enough reasonably flat back-shore area to permit the periodic movement of the machine. Once on the beach, the combustor would operate as a self-sufficient plant with the exception of fuel supply. Auxiliary firing would be accomplished using diesel oil, which would also fuel a motor/generator set. The latter would be rated at 45 hp for a 10 tph combustor. Combustor throughput has been very conservatively set and may well prove to be low by a factor as high as three. Establishment of this parameter was based on a thermodynamic analysis, in which several conservative assumptions had to be introduced.

Air Pollutants: When properly fired, particulate emissions from oil are quite low. In tests conducted essentially smokeless flames were obtained when oil/sand/water mixtures were combusted under process conditions. There is no reason why this should not also be the case in scaled-up systems. Observation of gaseous emissions was not undertaken. The insignificant levels involved should be acceptable considering the nature of the combustor's function.

Cost Factors: Cost estimations are based on 1970 data. While in service, the basic operation and maintenance costs of the combustor are relatively low. Based on an 8 hour day operation, the cleaning cost would come to $0.54/ton if the input sand contained 6 wt % oil and 6 wt % moisture. If sand of the same moisture but containing no oil were fired, the cost would increase to $2.78/ton. A number of other cost factors must also be considered. Most, however, are dependent on variables that cannot be readily fixed. Capital cost recovery, for example, must be set on the basis of some arbitrary utilization rate to determine chargeback. For example, if it were assumed that a 10 tph unit would be used only 30 working days per year, the user would be expected to pay $5.00 for each ton of sand cleaned. This would allow the owner to annualize his capital costs at a rate of return of only 8%.

Transportation costs have also been found to be highly variable. These will depend on the haul distance, the method of drayage, the need for exclusive vehicle privileges, and the availability of suitable unloading equipment and crews at the point of delivery. As a generalization, overall truck transportation costs would probably run about $25.00 for each hour the carrier is away from its terminal.

A final factor is the cost of operational support furnished by other personnel and equipment working at the site. It is felt, however, that because of the functions normally performed at a spill site, that no additional equipment or manpower would actually be required.